Feed-in Tariffs

Feed-in Tariffs
Accelerating the Deployment of Renewable Energy

Miguel Mendonça

London • Sterling, VA

First published by Earthscan in the UK and USA in 2007

ISBN-13: 978-1-84407-466-2

Typeset by MapSet Ltd, Gateshead, UK
Printed and bound in the UK by Cromwell Press, Trowbridge
Cover design by Nick Shah

For a full list of publications please contact:

Earthscan
8–12 Camden High Street
London, NW1 0JH, UK
Tel: +44 (0)20 7387 8558
Fax: +44 (0)20 7387 8998
Email: earthinfo@earthscan.co.uk
Web: **www.earthscan.co.uk**

22883 Quicksilver Drive, Sterling, VA 20166-2012, USA

Earthscan is an imprint of James and James (Science Publishers) Ltd and publishes in
association with the International Institute for Environment and Development

A catalogue record for this book is available from the British Library

Library of Congress Cataloging-in-Publication Data

Mendonça, Miguel.
  Feed-in Tariffs : accelerating the deployment of renewable energy / by Miguel
Mendonça.
     p. cm.
  Includes bibliographical references.
  ISBN-13: 978-1-84407-466-2 (hardback)
  ISBN-10: 1-84407-466-8 (hardback)
  1. Energy policy–Enironmental aspects. 2. Energy industries–Prioces–Government
policy. 3. Renewable energy sources–Government policy. 4. Energy
development–Environmental aspects–Government policy. I. Title.
  HD9502.A2M46 2007
  333.793'23–dc22

                                                           2007004107

Contents

PART 1 RENEWABLE ENERGY: BARRIERS AND SUPPORT SCHEMES

PART 2 POLICIES AROUND THE WORLD

PART 3 IMPLEMENTATION AND THE FUTURE

List of Figures, Tables and Box

FIGURES

TABLES

BOX

Acknowledgements

This text could not have been produced without drawing on the excellent work of the world's experts in the field of environmental policy, and their kind and generous assistance throughout the research process. In addition, many individuals and organizations have provided further valuable assistance and inspiration. To name just a few, I would like to extend my thanks and appreciation to Professor Herbert Girardet, Reinhard Hübner, Dr Colin Sage, Jakob von Uexküll, Daniel Bristow, Alistair Whitby, Tamsine Green and the staff at Earthscan, the staff, donors, advisers and councillors of the World Future Council (WFC), the people of Hamburg, the WFC expert panel members: Dr Janet Sawin (Worldwatch Institute), Paul Gipe (wind-works.org), Dr Hermann Scheer (World Council for Renewable Energy, Eurosolar, World Future Council), Dr Mario Ragwitz (Fraunhofer Institute), Dr Eric Martinot (Tsinghua University, REN21), José Etcheverry (David Suzuki Foundation), Craig Morris (petiteplanete.org) and Dr David Toke (University of Birmingham); Scott Sklar, Dr Peter Baz, Heiko Stubner, Dr Tetsunari Iida (ISEP), Dr Lutz Metz, Professor K. Vala Ragnarsdottir, Emeritus Professor Peter Haggett, Hugo Lucas Porta, Uwe Busgen, Josep Puig, Claudia Grotz and Ralf Bischof at the German Wind Energy Association, Gerhard Stryi-Hipp, Dr Frithjof Staiss, Dr Reinhard Haas, Dr Hubert Aulich, Stephen Karekezi, Soelyla and Werner Behrschmidt, Marcus Morrell at Bigpicture.tv, the German Federal Environment Ministry, IDAE (Instituto para la Diversificación y el Ahorro de la Energía), the staff at Schumacher UK, The Schumacher Institute for Sustainable Systems, Eurosolar, Renewable Energy Access.com, Earthwire UK, The Feed-in Cooperation, EREC, REEEP and the E-parliament, the University College Cork Geography department, the Bristol University Earth Sciences department and also to the Mendonças, the Blachfords and the Applegates, Dexter Banks, Michael Wallis, Dennis Keogh, Phil Elsmore, Matthew Schlanker and Corrina Thornton.

Foreword

Hermann Scheer

I am very pleased to write the foreword to this important book. As a member of the World Future Council I am particularly pleased that we are dedicating our first international campaign to renewable energy feed-in legislation. I think it is true to say that in my country, Germany, this legislation has been implemented to great effect.

Our world is faced with an intertwined energy and climate crisis. We must act vigorously. The key question is: who should and can do the job and with what tools?

For 200 years, industrial civilization has relied on the combustion of abundant and cheap carbon fuels. But continued reliance has had perilous consequences. On the one hand there is the insecurity of relying on the world's most unstable region – the Middle East – compounded by the imminence of peak oil, growing scarcity and mounting prices. On the other, the potentially cataclysmic consequences of continuing to burn fossil fuels, as the evidence of accelerating climate change shows.

Many agree that renewable energy is the future. I would submit that we should make a rapid transition to renewable sources of energy and distributed, decentralized energy generation. This is a model that has been proven, technologically, commercially and politically, as has now been comprehensively demonstrated. The alternative of a return to nuclear power – which is once again being widely advocated – is compromised and illusory. Apart from many other problems – of safety, long-term supply of uranium and storage of nuclear waste – there is not enough time to make a large-scale switch to nuclear energy. Nature's deadlines are simply too tight.

The installation of a large-scale power plant takes 5 to 15 years whilst a wind turbine can be installed in one week. The main objection to renewable energy is intermittence of supplies. But the fact is that energy storage technology – such as pumped storage and compressed air power plants, and hybrid structures to enhance efficiency, such as hydropower or bioenergy hook-ups for sun and wind energy plants – are well developed. The possibility for rapidly increasing the proportion of renewables in the energy mix, and moving towards a situation where all energy is provided by renewables, has become entirely plausible.

The advantages of renewable energy are so clear and so overwhelming that resistance to them needs careful examination. There are many who still pay lip service and who say that the promotion of renewable energy should not be their job, but that of others. Based on my experiences everywhere in the last 30 years, my recommendations are: Don't leave the job only to the market. A fair market requires equal market conditions, but these currently do not exist: trillions in subsidies were – and still are – being spent on nuclear and fossil energies, directly and indirectly. As this book explains very eloquently, the key to a sustainable energy future is renewable energy feed-in legislation.

In my country wind power has been making particularly rapid progress on the back of this legislation, but I would argue that Photovoltaics (PV) is ultimately the most promising of all energy technologies, giving us the best option to overcome global energy crises. Currently, it only generates a small proportion of total renewable energy supply, much less than wind, hydropower or biomass. Photovoltaic energy could soon become the 'prima donna' of renewable energy technologies.

Solar energy radiation is the only primary source directly exploitable at every place on Earth. It therefore offers everyone free access to energy and, moreover, to electricity, that most modern and multifaceted form of energy services. Thus PV has the potential to facilitate energy freedom for everybody – free from discrimination, artificial national borders and administrative hurdles, and free from dependency on energy monopolies. It bears by far the biggest potential – larger than that of all other renewable energy sources, larger than anything to which fossil fuels and nuclear power could ever aspire.

All the ecological advantages associated with renewable energies must be clearly demonstrated. There is an energy mind-set extending even into the ranks of the energy economy that is subjectively based on nuclear and fossil energies, but not objectively on *all* energies. Many energy experts don't even notice this one-sided thinking any more. It developed during the period of absolute dominance of nuclear and fossil energy sources. The degree to which this thinking penetrates the scientific community can be seen very clearly in any energy statistics.

But the fact is that the calculations that give preference to fossil fuels and nuclear energy are always incomplete. They do not include the energy lost during mining, processing of the mined energy source, transport, storage, distribution over thousands of kilometres of power lines in some instances, or disposal. In other words, they consider isolated equipment and not systems, meaning their calculations are wrong. A correct calculation would immediately show that an efficient energy system is virtually impossible with fossil fuels.

The flow of an energy source determines political, economic, social and cultural developments. Our task is to transform the macro social economics into the micro incentives that will mobilize renewable energy technologies. It is crucially important to get the policy right, and feed-in legislation has been proved to be the most effective policy for this purpose.

This book by Miguel Mendonça, for the first time, traces its development in detail whilst also comparing it with other policies. It invites policy makers all over the world to adopt feed-in legislation for their countries or their regions. I am certain that this policy option, which has already been adopted in over 40 countries and regions, can help us to create a better and much safer world.

Dr Hermann Scheer is a member of the German Parliament, President of EUROSOLAR, General Chairman of the World Council for Renewable Energy (WCRE) and a member of the World Future Council.

Executive Summary

The acceleration of the transition from an energy system based on fossil fuels to one based on renewable energies is of paramount importance for a variety of reasons. Firstly, the health and survival of humanity is dependent on the health and survival of the planet's natural resources and systems, which are being degraded and disrupted, principally by the burning of fossil fuels. Secondly, energy independence and security of energy supply for all countries would increase overall global security and enhance the conditions for peace. Thirdly, the potential economic advantages of quickly developing existing renewables markets or creating new industries are enormous. These advantages are available to both developed and developing countries, to urban and rural areas through the variety of different renewable technologies.

These issues are set to increase drastically in importance over time, and scientific evidence continues to document the accelerating decay of the conditions for life on Earth. Renewable energy technologies can dramatically change the patterns of ecological destruction that have become our collective signature over the last few decades in particular, but they still require support to make them competitive, and also to spread in sufficient quantities.

In countries where renewables have been supported well, enormous environmental, social and economic gains have been made. First movers have taken huge market shares in terms of technology development, manufacture and supply. Carbon dioxide emissions have been saved in vast quantities. Hundreds of thousands of jobs have been created. But this is only the beginning. The sector is growing rapidly, leaving enormous scope for others to take advantage of the opportunities in all parts of the supply chain. With so many drivers for market development then, why has it been so slow? This is essentially because it has been prevented by the various barriers to renewables that have at times been consciously erected.

BARRIERS

Barriers to renewable energy include: large public subsidies for fossil and nuclear fuels, in conjunction with a lack of assessment over their future costs; opposition from utility companies who are intolerant of competition from other (often smaller) investors; government preference for large, centralized solutions; lack of administrative experience with renewables; investment risk and uncertainty; and regulatory and institutional issues. These and other factors combine to distort the conditions for market access for renewables. Additional barriers in many developing countries include long travel distances to remote areas, poor transport and communication infrastructure, lack of trained personnel and low literacy rates. In addition, the perceived risk of investing in renewable energy projects in developing countries is high due to uncertainties about political, regulatory and market stability.

In view of vital renewables development being slowed by such barriers, measures are required to overcome them. The most successful policy instrument yet devised for speeding the comparatively low-cost deployment of renewable energy technologies is the feed-in tariff (FIT) model. This is essentially a 'pricing law', where the rates paid to producers of electricity from renewable sources are set in law, and calculated to provide sufficient profitability for the investment. There is usually a set period over which this payment will be received.

The superiority of the FIT instrument has been determined through experiences predominantly in Europe (where it is by far the dominant model), but also increasingly in other regions of the world. Many empirical studies, on which this text is in part based, have shown that in comparison to other instruments, as described in later chapters, this model has proven to be the most effective at overcoming barriers and rapidly generating the benefits of renewables for citizens, businesses and governments.

One of the other major barriers to renewables, one that is often underrated, concerns the myths about renewables, which form mental barriers. These myths often centre around the suggestion that countries cannot get more than a small percentage of their energy from renewables. However, if the world's fossil and nuclear fuels were to disappear tomorrow, it is certain that renewables would quite quickly come to provide our energy needs, through the new-found will of governments and utilities in particular. Even if continuing to use conventional energy were not so dangerous, it is nevertheless running low in various sectors as demand continues to spiral upwards. It is clear that if we are to continue to live on this planet, renewables – and making a science of energy efficiency – are the only long-term options we have, and that future can only be made more certain by rapidly reducing fossil fuel use.

As these realities become increasingly apparent at the political level, prompting more effective policy initiatives, technological and technical development will only accelerate. Experience is guiding development, and the commercial opportunities are being discussed relentlessly at conferences and in board rooms around the world. It will become the biggest global market, removing the technical, financial and political blockages to the overhaul of our global energy infrastructure and supply.

Poorly designed or inconsistent policies can actually harm renewables expansion, but countries like Germany, Spain and Denmark have demonstrated the possible outcomes of well-designed FITs, which are now found (at the time of writing) in 41 countries, states and provinces around the world (with various other countries discussing their introduction). Denmark generates 20 per cent of its electricity from wind power already, a figure that is still a distant goal for many countries. Much of this is due to a combination of the FIT that they had in place until 2000, and a culture of cooperation and environmental care – a social factor which is not to be underestimated.

Spain has a very effective FIT, which has helped to make them one of the world's top renewables manufacturers, with highly successful companies. They have set themselves a target of producing 30 per cent of their electricity from renewables by 2010, and are presently at 16.2 per cent and climbing. Germany has gone from having no real renewables market to becoming a world leader in renewable energy capacity, technology, manufacture and policy, with a new, multibillion-euro industry and over 214,000 new jobs – within just 15 years of implementing a feed-in law. As well as success with renewables in business terms, Germany has also helped to lead the way for other national, state and provincial governments.

OTHER COUNTRIES AND SUPPORT SCHEMES

The US, UK and Japan have all taken a different route in terms of policy, and none has taken advantage of FITs so far, instead choosing the quota model. Instead of setting the price for renewable energy (RE), quota systems set the quantity to be delivered. Utilities must provide a set amount of green electricity, for which they receive certificates. These certificates can then be traded or sold for extra income. Empirical studies have confirmed that they are producing less RE deployment, at a higher cost than FITs.

These countries have huge market opportunities for renewables, and contribute enormously to the emission of carbon dioxide and other greenhouse gases. The US is still the world's biggest polluter, a non-signatory of the Kyoto Protocol, and has so far relied on a number of proactive cities and states to take action on climate change. The federal government has steadfastly refused emissions caps, and has raised the ire of the global community by refusing to take responsibility on the issue. Their installed capacity of renewables remains minor in comparison with the potential. The enormous land mass and coastlines provide potential for every technology, including geothermal. Some experimenting is being done with 'hybrid' support schemes for renewables, which blend elements of feed-in and quota models. The Renewables Portfolio Standards (RPS) quota system that is employed in around half the US states has been criticized for various reasons, although it does at least represent a start in developing support schemes for renewables.

The UK has dwindling North Sea gas and oil reserves, and relies significantly on uncertain supplies from Russia. Wholesale gas prices have risen sharply over the last few years, and fuel poverty is becoming an increasing issue in poorer levels of society. On the other hand, the UK has Europe's greatest wind resources, opportunities in tidal technology and also great wave energy potential. However, examinations of the UK's quota system design have found it to be expensive and inefficient.

Japan has a reputation for energy efficiency, is a pioneering nation in terms of solar technology especially, and hosted the negotiations that led to the establishment of the Kyoto treaty. Unfortunately, a variety of conditions has led to the country being, itself, likely to miss the emissions reduction targets that it signed up to in the treaty. RE potential is significant, with geothermal being a promising area in this volcanic region. Again, the quota system at work has been considered rather poorly designed and implemented.

In short, the potential, as well as the necessity, for renewables in these countries is huge, yet the support schemes in place are, in terms of scale and technological variation, not delivering what a well-implemented feed-in system could. Among other criticisms, there is often a lack of support across all technologies, as quota systems tend to result in only the cheapest technologies being invested in first. Also, the greater investment risk inherent in quota systems in general acts to restrict the size of investor that can participate. While this is no object to large utility companies, householders cannot then make investment in residential photovoltaics (PV) cost-effective without further incentives.

Although 'developing countries' is a highly contested term, there are a number of socio-cultural, economic and geographical or environmental challenges facing most developing countries that make talking of them as a group useful. The most relevant of these for FITs is the degree to which the country is covered by a national grid. Experience suggests that off-grid areas are often better served by renewables designed to meet the energy needs of the local population, rather than by costly and technically challenging grid extensions. For grid-connected areas, the examples of Mauritius and China show the flexibility of the FIT model, and demonstrate both how FITs can encourage a growth in renewables, and how country-specific circumstances will help determine its success. The unifying message for developing countries is that they must develop a market for renewables, either through recognizing the synergies between development efforts and programmes to change energy use, or through creating the right policy environment to encourage investment – part of which should be the introduction of FITs where applicable.

POLICY DESIGN AND SUPPORT

Policy design for any support scheme is of critical importance. The research literature is emphatic on this, and agrees on a list of common elements that should be considered when designing legislation. European Union (EU) Member States have a variety of design options within their schemes in response to their own resources, renewables targets, electricity system and planning laws. No two feed-in laws are quite alike for this reason.

Key elements for policy design

- Provide tariffs for all potential developers, from householders to utility companies.
- Provide financial security – ensure that tariffs are high enough to cover costs, encourage development and provide a reasonable return on investment. Guarantee tariffs for a long enough time period to ensure high enough rate of return. The long-term certainty that results from guaranteed prices over perhaps 20 years allows companies to invest in technology, to train staff, and establish other services and resources with a longer-term perspective. This certainty also makes it easier to obtain financing, as banks and other investors are assured a guaranteed rate of return over a specified period of time. In fact, even banks in Germany lobbied the Bundestag for a continuation of feed-in laws in 2000.
- Remove barriers to grid connection. Guaranteed access to, and strategic development of the national grid are essential elements in developing investor confidence.
- Appropriate and streamlined administrative and application processes. In general, FITs are easier to administer and enforce than quota systems, and are more transparent. As with quota systems, policy makers are required to establish targets and timetables, and to determine which technologies are qualified (type and scale).
- Public acceptance. This is likely to be engendered by low costs to end consumers in particular, and demonstrable policy successes (Sawin, 2004).

In addition, some technologies may need extra support that should not necessarily be provided within a feed-in law. Innovative technologies that are still under development may therefore be better served with tax incentives or soft loans, for example. The same may apply to stand-alone projects.

OUTCOMES

In terms of the outcomes of the legislation, the figures from Germany are particularly telling. This is due to the decade and a half of development their feed-in laws have enjoyed, and sustained commitment from government and the public. According to the Federal Environment Ministry, the use of renewables prevented the emission of 83 million tonnes of CO_2 in 2005. In 2006 the share of total electricity consumption in Germany from renewables rose to 11.8 per cent, and German renewables companies had a combined turnover of €21.6 billion. The sector currently employs over 214,000 people (more than the number employed in the conventional energy industries), and expects this to rise to 500,000 by 2020. By that year, investments of €200 billion are anticipated. Germany, Denmark and Spain are among the world's leaders in renewable energy technology, taking enormous market shares between them. They have all built on the opportunities afforded by sudden and rapid market development created through their FITs. The lessons are clear, whether your interests are based on our moral duty to prevent the destruction of life, on the desire to create jobs and industry, or to make your country, state or province secure in energy supply.

Introduction

Solutions. That is where the debate must move now, beyond the reiterations and relentless updates concerning the onrushing climate catastrophe, the collapse of a civilization starved of oil and the mass extinction of species that we have set in motion. Regardless of what motivates us to alter our course, we need to identify and implement the right solutions quickly.

Where better to start than with proven solutions? Why not simply communicate with one another, determine who has had the best results with the way they do things, and then assess how this might be replicated elsewhere? Thankfully, this is being done to an extent, and in the field of environmental policy one model stands out above all others. This text introduces the concept of the FIT model to policy makers, non-governmental organizations (NGOs), environmental campaigners, the financial community, researchers, academics, students and other interested parties. The aim of both the book and the policy it discusses is to assist in the process of accelerating the deployment of renewable energy to help achieve, in no particular order: increased global energy security; reduced carbon emissions; job creation; industry creation and development; and access to energy.

The book explores various aspects of FITs, and shows how they have been implemented in different countries and why they should be strongly considered by policy makers. The performance of alternative support schemes is also discussed. This work shows what can be learnt from the successes and failures from around the world, what the basic considerations are for designing successful support schemes, what opportunities are offered by an effective scheme, and why renewables are ready, today, to help speed us away from fossil fuel dependency and its repercussions.

Having drawn on many comparison studies of the different renewable energy support schemes, it is the overwhelming conclusion of the world's leading researchers in this area of policy that FITs – if well designed and implemented, and in concert with complementary programmes – give rise to the fastest, lowest-cost deployment of renewable energy.

They have succeeded in making domestic solar PV cost-effective, and halving payback periods. Without support, rooftop PV would not be an option for most householders at present. With a sufficient tariff, everyone with a roof can feed energy into the grid through their solar panels when they produce more energy than they use – and get paid for it. A simple, elegant method of bringing green electricity into the mainstream.

The transition from conventional energy to renewable energy is a no-brainer. If we stick doggedly with fossil fuels, we lose. If we try to wait until they run out, and continue to simply tinker with renewables, we lose. Only by getting our most basic need right, by transitioning to using energy sources that are free, limitless, and are environmentally, socially and geopolitically benign can we win-win-win. To continue down the present path, with much talk and little action, while the climate is pushed closer every minute to a catastrophic tipping point, is nothing short of a crime against present and especially future generations.

Therefore, this text shows part of the way forward. It principally examines the German and Spanish feed-in systems, and the design and development of efficient, cost-effective methods of accelerating the deployment of renewable energy. Energy efficiency, the critical complement to renewables, is addressed briefly. The different design options for feed-in systems are highlighted, with examples from EU Member States. The alternative support scheme systems in use around the world are critically assessed, and examples of how FITs have been introduced in developing countries are shown. Many examples of techno-logical advances that can quickly bring down the costs of RE are shown, with a view to exploding the rather worn-out but dogged myth that renewables are expensive, limited and will only ever be a sideshow to the eternal burning of fossil fuels.

Feed-in laws are just one element of the transition, which is but one aspect of ecological modernization. The transition phase contains many more areas that are beyond the scope of this introduction to FITs, including: providing renewables support for other sectors such as transport and heating; the array of considerations for the transition in different countries (including differing economic models); the technicalities of integrating renewables into the grid; the Kyoto Protocol, and in particular the Clean Development Mechanism (CDM) and Joint Implementation (JI); emissions trading; carbon capping; a political economy analysis of energy policy decision making; ecological tax reform; peak oil; issues related to the emerging economies of China, India and Brazil; an exploration of the dangers and inadequacies of fossil fuels and nuclear energy (see the work of Hermann Scheer and Craig Morris); liberalization of energy markets; and the creation of an international treaty on accelerated deployment of renewables.

Above all, the central message within this rather dry subject of policy is that the transition to renewables is inevitable, morally and ethically imperative and must be accelerated massively in order to reduce the risk of sending the climate system into a chaos period that critically disrupts our modern infra-structure and economic systems, and ultimately leads to a mass human, plant and animal extinction. This text is neither political statement nor industry lobbying, but a call to all decision makers to take action on multiple fronts by exploring this proven mechanism for achieving sustainability goals, including domestic energy security and industry creation. Empirically, there is no better environmental policy in the world than well-designed and complemented feed-in legislation for triggering rapid, low-cost renewable energy deployment.

List of Acronyms and Abbreviations

AET	average electricity tariff
BEDP	Bagasse Energy Development Programme
BMU	Federal Ministry of the Environment, Nature Conservation and Nuclear Safety
BMWi	Federal Ministry of Economics and Technology
BTL	biomass-to-liquid
CAP	Common Agricultural Policy
CCT	compulsory competitive tendering
CDM	Clean Development Mechanism
CEB	Central Electricity Board
CFL	compact fluorescent lamp
CHP	combined heat and power
CPUC	California Public Utilities Commission
CSI	California Solar Initiative
CSP	concentrated solar thermal power
CST	concentrating solar thermal
DOE	Department of Energy
DSIRE	Database of State Incentives for Renewables and Efficiency
EEG	Erneuerbare-Energien-Gesetz (Renewable Energy Sources Act)
EREC	European Renewable Energy Council
ERP	Emerging Renewables Program
EU	European Union
FIC	Feed-in Cooperation
FIT	feed-in tariff
GDP	gross domestic product
GEF	Global Environment Facility
GHG	greenhouse gas
GW	gigawatt
HEPCO	Hokkaido Electric Power Company
IEA	International Energy Agency
IRENA	International Renewable Energy Agency
ISEP	Institute for Sustainable Energy Policies
IOU	investor owned utility

IPO	initial public offering
kW	kilowatt
kWh	kilowatt hour
METI	Ministry of Economy, Trade and Industry
MSW	municipal solid waste
MW	megawatt
MWh	megawatt hour
NDRC	National Development and Reform Commission
NGO	non-governmental organization
PBI	performance-based incentive
PTC	production tax credit
PPA	power purchase agreement
PURPA	Public Utility Regulatory Policies Act
PV	photovoltaics
RD	Royal Decree
R&D	research & development
RD&D	research, development & demonstration
RE	renewable energy
REEEP	Renewable Energy and Energy Efficiency Partnership
REL	Renewable Energy Law
REN21	Renewable Energy Policy Network for the 21st Century
RES	renewable energy sources
RES-E	renewable energy sources of electricity
RETD	Renewable Energy Technology Deployment
RO	Renewables Obligation
RPS	renewables portfolio standards
SHS	solar home systems
SIE	Sugar Industry Efficiency Act
SSPD	Sugar Sector Package Deal Act
StREG	Stromeinspeisungsgesetz (Electricity Feed Law)
TLP	tension leg platform
TW	terawatt
TWh	terawatt hour
UCS	Union of Concerned Scientists
VDEW	German Electricity Association
WHO	World Health Organization
Wpeak	Watt-peak

Part 1

Renewable Energy Support Schemes

1

Barriers to Renewable Energy

When addressing the reasons behind the hitherto slow transition to a 'solar economy' – one based on renewable energy (RE) as opposed to fossil fuels or nuclear – there are several main categories of barriers to address. These are chiefly cost, administration, technical and legal. Policies and programmes are required to support RE due to the existence of these barriers in the short term, but there is also a need to eradicate such barriers completely over time. These barriers are arguably dominated by the large subsidies that are still supplied to the fossil fuel and nuclear industries, even after many decades of support. Feed-in laws help offset these subsidies, but reducing them independently has proven politically difficult, especially when dealing with elected officials who represent coal states, for example. This tends to result in increasing RE subsidies rather than reducing subsidies for conventional energy.

This chapter examines these issues, and draws primarily on Beck and Martinot (2004) and Sawin (2004). These barriers are primarily considered for developed countries, but the barriers for developing countries are referred to briefly in Chapter 8.

COSTS AND PRICING

- Subsidies for competing fuels. Large public subsidies for all energy forms, both implicit and explicit, can distort investment cost decisions. The World Bank and International Energy Agency estimate global annual subsidies for fossil fuels to be in the range of US$100–200 billion (please note that dollars and cents are US throughout unless otherwise stated), although such figures are very difficult to assess (by comparison, the world spends around $1 trillion annually on purchases of fossil fuels). Public subsidies can take many forms: direct budgetary transfers, tax incentives, research and development (R&D) spending, liability insurance, leases, land rights of way, waste disposal and guarantees to mitigate project financing or fuel price risks.
- High initial capital costs. Even though lower fuel and operating costs may make renewable energy cost-competitive on a life cycle basis, higher initial

capital costs can mean that RE provides less generation capacity per initial dollar invested than conventional energy sources. Thus, RE investments generally require higher amounts of financing for the same capacity. Depending on the circumstances, capital markets may demand a premium in lending rates for financing RE projects because more capital is being risked up front than in conventional energy projects. RE technologies may also face high taxes and import duties. These duties may exacerbate the high first-cost considerations relative to other technologies and fuels.

- Difficulty of fuel price risk assessment. Risks associated with fluctuations in future fossil fuel prices may not be quantitatively considered in decisions about new power generation capacity because these risks are inherently difficult to assess. Historically, future fuel price risk has not been considered an important factor because future fossil fuel prices have been assumed to be relatively stable or moderately increasing. Thus, risks of severe fluctuations are often ignored. However, greater geopolitical uncertainties and energy market deregulation have bred new awareness concerning future fuel price risks. RE technologies (with the exception of biomass) avoid fuel costs and fuel price risk. This 'risk-reduction premium', is, however, often missing from economic comparisons and analytical tools because it is difficult to quantify. Furthermore, for some regulated utilities, fuel costs are factored into regulated power rates, so that consumers rather than utilities bear the burden of fuel price risks, and utility investment decisions are made without considering fuel price risk.

- Unfavourable power pricing rules. RE sources feeding into an electric power grid may not receive full credit for the value of their power. Two factors are at work here. Firstly, RE generated on distribution networks near final consumers rather than at centralized generation facilities may not require transmission and distribution (i.e. would displace power coming from a transmission line into a node of a distribution network). But utilities may pay only wholesale rates for the power, as if the generation was located far from final consumers and required transmission and distribution. Thus, the 'locational' value of the power is not captured by the producer. Secondly, RE is often an 'intermittent' source whose output level depends on the resource (i.e. wind and sun) and cannot yet be entirely controlled. Utilities cannot count on the power at any given time and may lower prices for it. Lower prices take two common forms: (i) a zero price for the 'capacity value' of the generation (utility pays only for the 'energy value') and (ii) an average price paid at peak times (when power is more valuable) which is lower than the value of the power to the utility, even though the renewable energy output may directly correspond with peak demand times and thus should be valued at peak prices.

- Transaction costs. RE projects are typically smaller than conventional energy projects. Projects may require additional information not readily available, or may require additional time or attention to financing or permitting because of unfamiliarity with the technologies or uncertainties over performance. For these reasons, the transaction costs of RE projects – including resource assessment, siting, permitting, planning, developing project proposals, assembling financing packages and negotiating power-purchase contracts with utilities – may be much larger on a per-kW capacity basis than for conventional power plants. Higher transaction costs are not necessarily an economic distortion in the same way as some other barriers, but simply make renewables more expensive. However, in practice some transaction costs may be unnecessarily high – for example, overly burdensome utility interconnection requirements and high utility fees for engineering reviews and inspection.

- Environmental externalities. The environmental impacts of fossil fuels often result in real costs to society, in terms of human health (e.g. loss of work days, health care costs), infrastructure decay (e.g. from acid rain), declines in forests and fisheries and, perhaps ultimately, the costs associated with climate change. 'Dollar' costs of environmental externalities are difficult to evaluate and depend on assumptions that can be subject to wide interpretation and discretion. Although environmental impacts and associated dollar costs are often included in economic comparisons between renewable and conventional energy, investors rarely include such environmental costs in the bottom line used to make decisions. If externalities were factored in, some renewables, particularly wind power, would already be cheaper than conventional energy sources.

LEGAL AND REGULATORY

- Lack of legal framework for independent power producers. In many countries, power utilities still control a monopoly on electricity production and distribution. In the absence of a legal framework, independent power producers may not be able to invest in RE facilities and sell power to the utility or to third parties under so-called 'power purchase agreements'. Or utilities may negotiate power purchase agreements on an individual ad hoc basis, making it difficult for project developers to plan and finance projects on the basis of known and consistent rules.
- Restrictions on siting and construction. Wind turbines, rooftop solar hot-water heaters, photovoltaic (PV) installations and biomass combustion facilities may all encounter building restrictions based upon height, aesthetics, noise or safety, particularly in urban areas. Wind turbines have faced specific environmental concerns related to siting along migratory bird paths and coastal areas. Urban planning departments or building inspectors may be unfamiliar with RE technologies and may not have established procedures for dealing with siting and permitting. Competition for land use with agricultural, recreational, scenic or development interests can also occur.
- Transmission access. Utilities may not allow favourable transmission access to RE producers, or may charge high prices for transmission access.

Transmission access is necessary because some RE resources, like windy sites and biomass fuels, may be located far from population centres. Transmission or distribution access is also necessary for direct third-party sales between the RE producer and a final consumer. New transmission access to remote RE sites may be blocked by transmission access rulings or right of way disputes.

- Utility interconnection requirements. Individual home or commercial systems connected to utility grids can face burdensome, inconsistent, or unclear utility interconnection requirements. Lack of uniform requirements can add to transaction costs. Safety and power-quality risk from non-utility generation is a legitimate concern of utilities, but a utility may tend to set interconnection requirements that go beyond what is necessary or practical for small producers, in the absence of any incentive to set more reasonable but still technically sound requirements. In turn, the transaction costs of hiring legal and technical experts to understand and comply with interconnection requirements may be significant. Policies that create sound and uniform interconnection standards can reduce interconnection hurdles and costs.

- Liability insurance requirements. Small power generators may face excessive requirements for liability insurance, particularly generators using 'net metering' provisions. Net metering allows customers to use their own generation, from PVs usually, to offset their consumption by allowing their electric meters to run backwards when they generate electricity in excess of their demand. This offset means that customers receive retail prices for the excess electricity they generate. The phenomenon of 'islanding', which occurs when a self-generator continues to feed power into the grid when power flow from the central utility source has been interrupted, can result in serious injury or death to utility repair crews. Although proper equipment standards can prevent islanding, liability is still an issue. Several US states have prohibited utilities from requiring additional insurance beyond normal homeowner liability coverage as part of net metering statutes.

MARKET PERFORMANCE

- Lack of access to credit. Consumers or project developers may lack access to credit to purchase or invest in RE because of lack of collateral, poor credit ratings, or distorted capital markets. In rural areas, 'microcredit' lending for household-scale RE systems may not exist. Available loan terms may be too short, relative to the equipment or investment lifetime. In some countries, power project developers have difficulty obtaining bank financing because of uncertainty as to whether utilities will continue to honour long-term power purchase agreements to buy the power.

- Perceived technology performance uncertainty and risk. Proven, cost-effective technologies may still be perceived as risky if there is little experience with them in a new application or region. The lack of visible installations and familiarity with RE technologies can lead to perceptions of greater technical risk than for conventional energy sources. These perceptions may increase required rates of return, result in less capital availability, or place more stringent requirements on technology selection and resource assessment. 'Lack of utility acceptance' is a phrase used to describe the historical biases and prejudices on the part of traditional electric power utilities. Utilities may be hesitant to develop, acquire and maintain unfamiliar technologies, or give them proper attention in planning frameworks. Finally, prejudice may exist because of poor past performance that is out of step with current performance norms.

- Lack of technical or commercial skills and information. Smooth market function requires low-cost access to good information and the requisite skills for all concerned. However, in specific markets, skilled personnel who can install, operate and maintain RE technologies may not exist in large numbers. Project developers may lack sufficient technical, financial and business development skills. Consumers, managers, engineers, architects, lenders or planners may lack information about RE technology characteristics, economic and financial costs and benefits, geographical resources, operating experience, maintenance requirements, sources of finance and installation services. The lack of skills and information may increase perceived uncertainties and interfere with decision making.

2

Renewable Energy Support Schemes

Many studies have compared the different mechanics, merits and performance of the different RE support schemes, and have concluded that feed-in laws have produced the quickest, lowest-cost deployment of renewable technologies in countries that have implemented them well. This chapter looks at the main support scheme models in use around the world today, and assesses their effectiveness.

THE FEED-IN MODEL

Feed-in laws are still a predominantly European phenomenon, but as the Renewable Energy Policy Network for the 21st Century (REN21) Global Status Report 2006 Update shows, it is now in place in 41 countries, states and provinces, including parts of India and Canada. It is also, in various forms, either already in place, or being discussed in parts of Africa, Asia, the Americas and Australia. Table 2.1 shows the uptake of the policy around the world up to the early part of 2006.

The basic feed-in model could be considered a 'pricing law', under which producers of renewable energy are paid a set rate for their electricity, usually differentiated according to the technology used and size of the installation. The rate should be scientifically calculated to ensure profitable operation is guaranteed. The period for which that rate is received should also be set in law, and should cover a significant proportion of the working life of the installation. Grid operators are obliged to provide priority access to the grid for RE installations.

The additional costs of these schemes are paid by suppliers in proportion to their sales volume and are passed through to the power consumers by way of a premium on the kilowatt hour (kWh) end-user price. In the best designs, the guaranteed periods are long, thus providing investment certainty. A variant of the feed-in tariff (FIT) scheme is the fixed premium mechanism currently

Table 2.1 *Cumulative number of countries/states/provinces enacting feed-in policies, 2005*

Year	Cumulative number	Countries/states/provinces added that year
1978	1	United States
1990	2	Germany
1991	3	Switzerland
1992	4	Italy
1993	6	Denmark, India
1994	8	Spain, Greece
1997	9	Sri Lanka
1998	10	Sweden
1999	13	Portugal, Norway, Slovenia
2000	14	Thailand
2001	16	France, Latvia
2002	20	Austria, Brazil, Czech Republic, Indonesia, Lithuania
2003	27	Cyprus, Estonia, Hungary, Korea, Slovak Republic, Maharashtra (India)
2004	33	Italy, Israel, Nicaragua, Prince Edward Island (Canada), Andhra Pradesh and Madhya Pradesh (India)
2005	40	Turkey, Washington (US), Ireland, China, India (Karnataka, Uttaranchal, Uttar Pradesh)
2006	41	Ontario (Canada)

Source: REN21, Global Status Report 2006 Update

implemented in Denmark and partially in Spain. Under this system, the government sets a fixed premium or an environmental bonus, paid above the normal or spot electricity price to RE generators. In the face of the established opposition from highly subsidized fossil fuel and nuclear industries (which the electricity network is designed to deal with), renewables are at an obvious disadvantage. Much of the extra costs for renewables result from these subsidies and other barriers, as discussed in the previous chapter.

Other design options, and differences in tariffs and other aspects of the law are discussed in the chapters on Germany and Spain, and in the chapter on design options (Chapters 4, 5 and 9 respectively).

Figure 2.1 shows the spread of support schemes in the 25 Member States of the European Union (EU-25), and illustrates the dominance of the feed-in instrument.

To give a clearer idea of why FITs are so desirable for the efficient advancement of RE, the main alternatives in use today, and their comparative shortcomings, will be briefly examined.

THE QUOTA MODEL

The quota system is used extensively in the US and to a small extent in Europe – notably by the UK and Sweden. While feed-in laws set the price and let the market determine capacity and generation, quota systems work in reverse. In

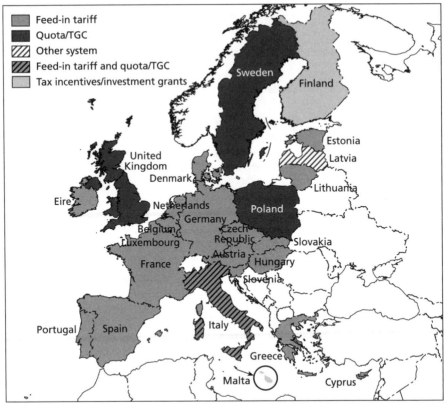

Source: Klein et al, 2006

Figure 2.1 *Currently applied schemes for the support of electricity from renewable energy sources in the EU-25 countries*

general, governments mandate a minimum share of capacity or (grid-connected) generation of electricity to come from renewable energy sources (RES). The share often increases over time, with a specific final target and end-date. The mandate can be placed on producers, distributors or consumers.

There are two main types of quota systems used today: obligation/certificate and tendering systems. The Renewables Portfolio Standard (RPS), widely used in the US, is in the former category. Under an RPS, a target is set for the minimum amount of capacity or generation that must come from renewables, which should increase over time. Investors and generators then determine how they will comply, in terms of the type of technology to be used (except in the case where specific targets are established by technology type), the developers to do business with, and the price and contract terms they will accept. At the end of the target period, electricity generators (or suppliers, depending on the policy design) must demonstrate, through the ownership of credits that they earn through transactions, that they are in compliance in order to avoid paying a penalty. Producers receive credit in the form of 'green certificates' for the renewable electricity they generate, which can be traded or sold, to serve as proof of meeting their legal obligation and to earn additional income. (Some countries have set floors and/or ceilings for the value that these certificates can achieve.) Those with a surplus of certificates can trade or sell them; those with too few can build their own renewable capacity, buy electricity from other renewable plants (which generally involves a bidding process), or buy credits from others. Once the system has been established, government involvement includes the certifying of credits, and compliance monitoring and enforcement (Sawin, 2004, p6).

Although the theory appears sound, and quota systems have produced increases in RE, the reality is that the system is in general significantly less cost-effective than the FIT system, and is flawed and inequitable in a number ways. This has been found in each country discussed in this book. A recent report by The Carbon Trust, a government-sponsored body, into the effectiveness of the UK quota system found the following:

> *The main pillar of the current renewable energy policy, the Renewables Obligation (RO), will cost consumers c.£14bn by 2020 and c.£18bn by 2027 (in present value terms). It is expected to result in renewables penetration at 7.6 per cent, 9.6 per cent and 10.1 per cent of generation by 2010, 2015 and 2020 respectively. This would mean renewable energy penetration would be only three-quarters towards the target for 2010, and only halfway towards achieving the 2020 aspiration of 20 per cent. Performance against these goals is held back in part by frictions such as planning and grid constraints, which are restricting installed capacity of the lowest cost technology (onshore wind), but also by the inefficiencies of the policy itself [emphasis added].*

> *The overall cost of installed renewable energy to consumers will be higher than necessary, given the current technology cost because the RO is inefficient in a number of ways. First, because the RO is designed to 'pull through' lowest cost technologies sequentially, it is not closing the funding gap for offshore wind fast enough to stimulate the necessary momentum for the 2015 timeframe. It is not succeeding in driving offshore wind installation and reductions in the offshore wind cost curve. Secondly, the RO (by design) passes regulatory risk to the private sector, which the private sector accordingly prices at a premium. This leads to leakage of the subsidy away from developers, as suppliers take a margin to deal with this risk and funding from financiers is therefore available on less favourable terms than it would otherwise be. There is wide and growing consensus that the RO needs to be adapted or changed; not doing so will introduce an element of political risk that may be very difficult to*

manage, associated with sustained high Renewable Obligation Certificate (ROC) prices and renewables delivery below Government targets. Without additional support for offshore wind, the development of offshore wind installations at scale in the UK (and very likely worldwide) will be held back; the UK renewable energy targets will be missed by a wide margin, carbon emission reduction targets will be harder to meet and an opportunity for renewable energy to become a meaningful component of the UK's energy mix may pass (Carbon Trust and LEK Consulting, 2006, p2).

Another 2006 report examining the RO found the following:

Another central objective of this research project was to investigate the effectiveness of the British Renewables Obligation (RO) in comparison with other types of financial support mechanisms for wind power and renewables. The RO is a mechanism involving creation of a market in green electricity certificates. In an article in Environment and Planning C *it was concluded that the RO did create conditions for a major expansion of renewable energy, but that a well-funded feed-in tariff of the sort organised in Spain, and [...] Germany, had many attractions. Feed-in tariffs involved prices for electricity from renewable generators being fixed by the Government. There is evidence that the British RO is less cost effective compared to feed-in tariffs and that its 'one price for all technologies' was not ideal for supporting offshore windfarms which are more expensive than most proposed onshore windfarms. The Government has proposed that the RO be banded. However there is no reason to believe that this reform will solve the problems inherent in the RO.*

A central problem with the RO as a financial support system is that there is considerable uncertainty about the future value of renewable obligation certificates (ROCs) and also of the electricity itself. This means that investors in wind power schemes will earn 'risk premia' for investing in schemes which have a lack of certainty of income flow, whilst in reality the prices of ROCs are high because of unfulfilled quotas. Attempts to control the value of ROCs within a banding arrangement will add to complexity and increase the relative attractiveness of a 'feed-in' tariff system which is simpler and also able to give an income stream, in £s per MWh, that takes account of the wind speeds on different sites.

A central financial issue is the funding of offshore wind power. Offshore wind power schemes are, in general, more costly than onshore schemes, and the problems of financing offshore projects are compounded on the relative lack of experience and thus confidence

*in such schemes compared to onshore projects. The RO does not
provide sufficient amounts or security of income to allow the
offshore wind power schemes which have already gained planning
consent, to be built. A 'feed-in' tariff system could improve security.*
(Toke and Marsh, 2006)

Although a detailed critique of all the support schemes and their variants is
beyond the scope of this book, the UK design shows clear failings in terms of
risk reduction, cost and speed of deployment. An important difference between
the two systems is that a quota system relies somewhat more on market forces.
Some believe that the market is the only tool that should be employed in any
commercial enterprise, and feed-in legislation is therefore less attractive, even
though both systems clearly create market distortions. The prevailing economic
ideology of a government may therefore influence their decision on which type
of policy to adopt. The danger here is that the vital goal of the rapid switch to
renewables, and all its benefits, may be lost due to this short-sighted decision
driver. Large utility companies are more able to bear the risks of the quota
system, thus reducing competition from decentralized generators. It is matter
of conjecture as to what influence the utilities have on government policy.

The US also uses a quota system (Renewable Portfolio Standards) in many
states. US renewables support measures are discussed in Chapter 6. The
findings there are very similar to those in the UK reports, and come together in
a common series of pros and cons, as summarized below.

Feed-in laws

Arguments for:

- To date, they have proven most successful at developing renewables markets
  and domestic industries, and achieving the associated social, economic,
  environmental and security benefits.
- Greater flexibility can be designed into the scheme to account for changes
  in technology and the marketplace.
- FITs encourage steady growth of small- and medium-scale producers.
- Low transaction costs.
- Ease of financing.
- Ease of entry.

Arguments against:

- If tariffs are not adjusted over time, consumers may pay unnecessarily high
  prices for renewable power. This can be addressed through monitoring.
- Can involve restraints on renewable energy trade due to domestic production requirements.

Quota systems

Arguments for:

- Promote least-cost projects – cheapest resources used first, which brings down costs early on.
- Theoretically provide certainty regarding future market share for renewables (often not true in practice).
- Perceived as being more compatible with open or traditional power markets.
- More likely to fully integrate renewables into electricity supply infrastructure.

Arguments against:

- High risks and low rewards for equipment manufacturers and project developers, which slows innovation.
- Price fluctuation in 'thin' markets, creating instability and gaming.
- Tend to favour large, centralized merchant plants and not suited for small investors due to greater investment risk.
- Concentrate development in areas with best resources, causing possible opposition to projects and missing many of the benefits associated with RE (jobs, economic development in rural areas, reductions in local pollution).
- Targets can set upper limits for development – there are no high profits to serve as incentives to install more than the mandated level because profitability exists only within the quota.
- Tends to create cycles of stop-and-go development.
- Complex in design, administration and enforcement, leading to a lack of transparency.
- High transaction costs.
- Lack flexibility – difficult to fine-tune or adjust in short term if situations change.

THE TENDERING SYSTEM

The third, if less commonly used support scheme, is the tendering system. This is administered by the government, and is a mechanism in which RE developers bid for power purchase agreements and/or access to a government-administered fund through a competitive bidding process. Regulators specify an amount of capacity or share of total electricity to be achieved, and the maximum price per kWh. Project developers then submit price bids for contracts. Governments set the desired level of generation from each resource, and the growth rates required over time. The criteria for evaluation are established prior to each round of bidding. There are sometimes separate tenders for different RE technologies. Generally, proposals from potential developers are accepted starting with the lowest bid and working upwards, until the level of capacity or generation required is achieved. Within each technology band, contracts (and the corresponding support) are awarded to the most competitive (i.e. lowest) bids. In some cases, governments will require separate bids for different technologies, so that solar PV is not competing against wind energy projects, for example. Electric utilities are often obliged to purchase the electricity at the price proposed by the winning bids (sometimes supported by a government fund). Those who win the bid are guaranteed their price for a specified period of time; on the flip side, electricity providers are obligated to purchase a certain amount of renewable electricity from winning producers at a premium price.

Tendering systems have been used in the UK, Ireland, France, the US and China. In comparison with RE deployment under FITs, tendering systems have performed poorly. Contributing factors include the intermittency of the tenders (creating uncertainty in the market) and the complexity of the procedures involved. It is also suggested that unrealistically low bids are often a result of government tendering processes, leading to the commitment of funds to projects that cannot be completed. This last point was rather painfully learnt in the UK, when the Conservative party introduced a policy for government contracts known as Compulsory Competitive Tendering, or CCT. This policy was eventually scrapped due to similar problems experienced by the tendering system, that is, that bids were too low to be sustainable, and so contract terms would be broken regularly, staff morale and work quality would be low. Finally, the incoming Labour government replaced CCT with a 'Best Value' system.

NET METERING

A variation on pricing laws, 'net metering' permits consumers to install small renewable systems at their homes or businesses and then to sell their excess electricity into the grid. This excess electricity must be purchased at wholesale market prices by the utility. In some cases, producers are paid for every kWh they feed into the grid; in other cases they receive credit only to the point where their production equals their consumption. This option is available in Japan, Thailand, Canada and at least 38 US states, including Texas and California. It is of benefit to electricity providers as well as system owners, particularly in the case of PV, because excess power generated during peaking times can improve system load factors and offset the need for new peak load generating plants.

Net metering differs from the access and pricing laws in Europe primarily in scale and implementation. Success in attracting new RE investments and capacity depends on limits set on participation (capacity caps, number of customers or share of peak demand); on the price paid, if any, for net excess generation; on the existence of grid connection standards; and on enforcement mechanisms. Without other financial incentives, net metering is not enough to advance market penetration. Neither California nor Texas saw much benefit from net metering for wind power, let alone for more costly renewables like solar PV, until other incentives were added to the mix. However, net metering might have a greater impact if private generators were to receive time-of-use rates for the electricity they put into the grid – particularly in the case of PV, which generate electricity at peak demand times when the value of their power is highest. Mandated targets or quotas, and net metering can be used simultaneously (Sawin, 2004, p5).

THE ALTERNATIVES ARE NOT WORKING

The conclusion among many researchers on this issue is that feed-in systems have many advantages over quota and tendering systems, and above all they get us where we need to go in terms of renewables deployment, faster and cheaper. It is questionable, on the face of it, that this model has not been seized upon by governments around the world in order to use the best energy and policy combination, renewables and feed-in laws, to rapidly develop a low-carbon energy supply infrastructure that cannot ever run out of free energy. Quota and tendering models have not been able to match this performance for the above reasons. One of the central problems, already stated, is that the alternatives do not support all technologies adequately, do not encourage geographic distribution of RE installations and do not allow householders to receive a sufficient rate of return to make investment worthwhile. These facts alone make them impractical in allowing a country to exploit all of its RE potential, which must be done in order to reach both renewables and emissions targets. The best tools are there to be used, and there are a number of organizations and ministries willing to advise and assist in the consultation, design and implementation process (see Chapter 10 on implementation).

In the case of onshore wind in EU Member States (as of 2004), installed capacity is clearly shown in Figure 2.2 to benefit most dramatically from FITs, as compared with any other instrument or incentive.

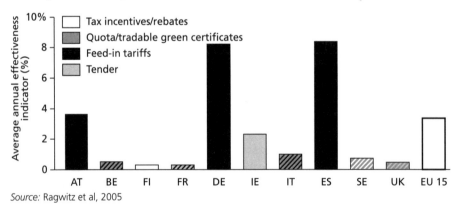

Source: Ragwitz et al, 2005

Figure 2.2 *Effectiveness indicator for wind onshore electricity in the period 1997–2004*

BOX 2.1 QUOTES FROM COMPARISON STUDIES OF RE SUPPORT SCHEMES

Renewable Tariffs have proven the most successful mechanism for stimulating investment in renewable electricity generation worldwide. Renewable Tariffs have resulted in more installed generating capacity and more robust competition among manufacturers and have stimulated more renewable technology development than any other policy mechanism. (Gipe, 2006, p1)

To date, feed-in – or pricing systems have been responsible for most of the additions in renewable energy capacity and generation, while also driving down costs through technology advancement and economies of scale, and developing domestic industries and jobs. Pricing systems, where well-implemented, have provided increased predictability and consistency in markets, which in turn has encouraged banks and other financial institutions to provide the capital required for investment, and has attracted private investment for R&D. (Sawin, 2004, p27)

The German EEG is more effective at increasing the share of renewables than the England and Wales RO because it reduces risk for RES generators more effectively. (Mitchell et al, 2003a, p4)

A well-designed (dynamic) FIT system provides a certain deployment of RES-e fastest and at lowest costs for society. (Haas et al, 2006, p27)

The EEG has been successful in deployment; in reducing risk; in developing an advocacy coalition; and in developing a new industry. The RO does not reduce risk (whether price, volume and market) which makes it difficult to obtain financing, so limiting new entrants and the development of an advocacy group. (Mitchell et al, 2003b, p2)

... until now so called renewable energy feed-in tariffs (REFITs) have shown the best effectiveness concerning the creation of new RES installations. (Bechberger and Reiche, 2006, p6)

The long-term price guarantee provided by the feed in tariff reduces regulatory and market risk. (Butler and Neuhoff, 2004, p31)

Feed-in tariffs have been proven to be successful elsewhere (Spain and Germany) in generating significant deployment of low-cost renewable energy. (The Carbon Trust and L.E.K Consulting, 2006, p3)

Feed-in tariffs are more efficient from a societal point of view compared to TGC [tradable green certificate] systems. In addition, they are useful to promote a more homogeneous distribution among different technologies by setting technology-specific guaranteed tariffs. The implementation of such a policy can support the long-term technology development of various RES-E options which are currently not cost-efficient. (Ragwitz et al, 2005, p38)

It is clear that the German 'feed-in tariff' has proved a highly flexible and manageable policy instrument. In contrast, the UK's Renewables Obligation (RO) has proved more costly but less productive. Moreover, the RO is making wind power progressively more expensive to the UK consumer at a time when degressive 'feed-in rates' are making it cheaper in Germany. (Szarka and Blühdorn, 2006, pv)

A stable feed-in tariff has clearly proven to be one of the most successful mechanisms to date for promoting large-scale wind energy markets that offer the stability necessary to attract local manufacturing. (Lewis and Wiser, 2005, p20)

The EU-15 figures lead to the conclusion that, when the feed-in tariffs are set correctly, the support scheme is able to start market development. The green certificate systems seem to need a secondary instrument (based on environmental benefits) for a real market effect. (European Commission, 2005, p33)

Quotes from other sources

Both sets of instruments have proved effective but existing experience favours price-based [feed-in] support mechanisms. Comparisons between deployment support through tradable quotas and feed-in tariff price support suggest that feed-in mechanisms achieve larger deployment at lower costs. Central to this is the assurance of long-term price guarantees. The German scheme ... provides legally guaranteed revenue streams for up to twenty years if the technology remains functional. Whilst recognising the importance of planning regimes for both PV and wind, the levels of deployment are much greater in the German scheme and the prices are lower than comparable tradable support mechanisms (though greater deployment increases the total cost in terms of the premium paid by consumers). Contrary to criticisms of the feed-in tariff, analysis suggests that competition is greater than in the UK Renewable Obligation Certificate scheme. (Stern, 2007, p366)

... it has clearly emerged that minimum price systems have proved more effective [than quota systems] in increasing clean energy capacity. (Bechberger and Reiche, 2005a, p16)

In stimulation of PV market growth, a feed-in tariff is the single most important and most successful driver, when applied correctly. (European Photovoltaic Industry Association, 2005, p3)

... only a model based on guaranteed feed-in tariffs enables a quick and broad implementation of renewable energy, better supports its technological development, as well as more efficiently promotes cost reduction. (Scheer, 2005a, p80)

3

Energy Efficiency: The Essential Partner of Renewables

Dan Bristow

Increasing energy efficiency is the natural complement to accelerating the uptake of renewables. Without this, the demand for energy is destined to increase indefinitely as countries strive to maintain economic growth – ultimately exceeding global productive capacity. Although the amount of energy used per unit of GDP has dropped for many developed countries, in spite of existing efficiency initiatives, energy use is projected to continue to grow over the next two decades. Implicit in this analysis is that the energy system as a whole needs to be addressed – not simply the processes of generation, transmission and distribution, but also the patterns of energy consumption.

The way we currently use energy is wasteful in the extreme, and is largely based on economic models based on out-dated assumptions about the abundance of primary fuels. Yet, despite seemingly countless examples of ways that we could maintain the status quo while using significantly less energy, change is painfully slow in coming. A clear example of this is the global electricity system, which wastes more than two-thirds of the primary energy that fuels it. With more than 60 per cent of this waste coming from heat loss in the production processes, simply switching to cogeneration (where the heat generated is captured and used to heat surrounding buildings) could reduce this waste to negligible levels. Similarly, there are significant energy savings to be made in end-use, which accounts for a further 13 per cent of the two-thirds lost (Greenpeace International and European Renewable Energy Council (EREC), 2007).

In the production of power, cogeneration or tri-generation plants typically achieve 90 per cent efficiency (International Energy Agency (IEA), 2004), providing electricity as well as heating (and cooling in tri-generation) for the surrounding buildings. While these generators run off the burning of carbon-based fuels, a shift to using biomass, and particularly waste biomass, would offer an efficient, carbon-neutral source of energy. There are a number of cities that have implemented cogeneration on a large scale. For example, in Woking in the UK, the council has halved its energy use and managed to reduce its emissions by 77 per cent since 1990 by introducing a network of such generators (Greenpeace, 2006a). In the city of Malmo, Sweden, they have gone a step further in the recent redevelopment of a section of the harbour. With their integrated energy system, this area of the city now produces 100 per cent of the energy it uses over the course of a year. The local system draws energy from a wind turbine, PV panels and solar thermal collectors attached to the buildings, an innovative aquifer storage system, and a district heating system. When excess energy is produced, it is 'exported' to the grid. When demand exceeds supply, the grid returns the favour (Ekostaden, 2005).

When combined with efficient building design, net metering can all but eliminate residential use of centrally produced energy. For example, the Passive House movement, which started in Germany, has helped to create buildings that are net *producers* of energy. Combining intelligent design with the latest technology, these houses show how, by making resource use a primary consideration, we can facilitate the kind of step-change necessary for tackling climate change. To qualify as a passive house, buildings have to achieve very high standards in terms of the energy used to heat and cool the space. Consequently, a passive house commonly has the highest standard of wall, window and joint insulation; it will typically have a greater number of windows on the southern facing aspect, thereby enabling greater capture of the sun's heat; similarly, it will use solar thermal collectors to heat water; and finally the ventilation system will use the warm outgoing air to heat the cooler incoming fresh air, or vice versa on warmer days (PassivHausUK, 2007). Once these efficiencies are achieved, adding microgeneration (e.g. a wind turbine or some PV panels) enables the minimal energy needs of the house to be met, or even exceeded, without drawing on centralized energy production processes. Existing housing stocks are also being made more efficient through a number of retrofit policies being pursued, in which funding is provided for improving insulation or replacing inefficient boilers, for example.

The energy efficiency of electrical appliances is another area where surprisingly large end-use savings can be made. Research by the IEA has suggested that within five years of their introduction, robust appliance efficiency policies could achieve a reduction in emissions equivalent to nearly 200 gas-fired power stations (IEA, 2003a, p15). As an example, the energy used by electrical goods that are left on standby is astounding. Although each item may use only between 0.5 and 10 watts while in standby mode, it is estimated that cumulatively these 'inactive' electrical goods, or 'energy vampires' account for roughly 1 per cent of the global carbon dioxide emissions (IEA, 2005). Adequately tackling this would involve introducing minimum efficiency standards, preferably at the international level.

Another candidate for international legislation is the incandescent light bulb. Almost 9 per cent of global primary energy consumption is used for artificial lighting. Although compact fluorescent lamps (CFLs) use between a quarter and a fifth of the energy used by incandescent light bulbs, they account for only 6 per cent of the lighting market (IEA, 2006). Light-emitting diode (LED) bulbs are even more efficient but are even more immature in the market. An international agreement to phase out the production of incandescent light bulbs would reverse this wasteful energy trend, and help reduce the associated emissions – approximately 8 per cent of the world's carbon dioxide emissions. China appears to be a step ahead, in that CFL bulbs are commonplace and incandescent bulbs are very difficult to find on sale. Australia has committed to phasing out the sale of incandescent bulbs by 2010.

This is just a small selection of the various ways in which energy efficiency can be improved in the residential and commercial sectors, but achieving these involves government action. FITs can be used to promote cogeneration/trigeneration and even microgeneration, but this has to be accompanied by complementary planning laws. An increase in end-use efficiency can largely be achieved through regulating at the national or international level for increases in the minimum efficiency standards of both buildings and electrical appliances.

In relation to energy use in the industrial and transport sectors, efficiency gains for both depend on the context, but there are some general observations to be made:

- Beyond efficiency gains to be made through technological advances (and the international spread of these), industrial processes need to be reoriented around 'closed system', 'circular economy' or 'industrial symbiosis' approaches, in which the waste produced through one process is either used as a resource for another, or recycled back into the same process. Recent legislative developments (particularly in the EU) that require producers to be responsible for the recycling of their products are a move towards this, but the concept has been more thoroughly put into practice in other parts of the world. In China, for example, the city of Guiyang is using the idea as the basis for its development and has agreed more than 20 projects designed around circular economy principles (People's Daily Online, 2005). In general a more widespread monitoring of energy and resource flows within and between particular industries would facilitate the identification of further resource efficiency gains.
- In simplified terms, a drive for energy efficiency in the transport sector involves improving mass transportation infrastructure, and limiting the unnecessary use of carbon-based forms of transport. In practice, domestically this involves careful planning of urban and industrial development. In America, for example, mass suburbanization combined with poor investment in public transport has meant that in many American cities, a large

proportion of the population are obliged to use private cars as their primary form of transport. In stark contrast to this is the city of Portland, where the development of an integrated transport system has encouraged a 65 per cent increase in the use of public transport, eliminating 62 million car journeys a year and helping reduce Portland's greenhouse gas (GHG) emissions to pre-1990 levels (*Newsnight*, 2005).

Finally, it is important not to overlook the 'Jevons Paradox', which suggests that increases in energy efficiency can result in increased energy consumption – the lowering of demand reduces the cost of energy, and thereby encourages greater consumption (Herring, 2006). One way of counteracting this involves rethinking the way utility companies make profits – if power companies no longer generated profits in relation to the throughput of energy, they would not be motivated to increase energy consumption. For example, if a utility company was contracted to supply the heating and lighting for a building for a certain period of time and at a fixed rate, they would have an incentive to reduce the energy consumption of that building because doing so would reduce their costs in the long term.

A related idea is that of 'Negawatts', which assumes the decoupling of profit and throughput, but adds to it the idea of creating a market in efficiency gains. For example, if a utility company predicts that growing demand will soon outstrip capacity, rather than building another power station or importing power from another utility, it could offer a tender for increasing efficiency (to a particular industry for example). Companies would then bid for the grant. If well managed, the utility would pay less to reduce demand than it would to increase capacity, and the company with the winning bid would recoup a decent proportion of its costs, while also making savings in reduced energy bills in the long term (Lovins, 1989).

One of the greatest advances in emissions reduction will come when an effective policy is agreed and implemented which decouples energy consumption from utility profits. This is perhaps one of the key approaches that can assist in the transition phase of moving from fossil fuels to renewables, and could help in changing attitudes towards energy production and consumption.

Part 2

Policies Around the World

4

Germany's Success

The State protects, on behalf of future generations, the natural foundations of life and animals...
Basic Law for the Federal Republic of Germany, Article 20 A

An energy usage is sustainable if the sufficient and permanent availability of suitable energy resources is assured, while at the same time limiting the detrimental effects of supplying, transporting, and using the energy.
German Federal Ministry for the Environment, Nature
Conservation and Nuclear Safety,
'Renewable Energies Innovations for the Future'

The history of FIT development in Germany shows how a gradual, trial and error approach to renewables legislation succeeded in bringing about substantial and ongoing environmental, social and economic benefits, with sustained social and political support being instrumental in this. The politics, as well as the policies, are of great importance in understanding the conditions for such success. The historical material is drawn chiefly from Kissel and Oeliger (2004), Mez (2004) and Lauber and Mez (2004). These papers, particularly Kissel and Oeliger's, explain the political processes behind the introduction of the law, and the battles that were fought to preserve the Act in a form that would ultimately be effective against the various barriers that confront renewables.

A BRIEF HISTORY OF ENERGY RESOURCES IN GERMANY

Germany's enormous endowment of coal and lignite made it easy to develop and sustain a huge energy supply from fossil fuels, aiding industrial growth and bolstering a powerful political force. The reliance on coal was ameliorated somewhat by the nuclear power industry which developed from the 1960s.

After the oil crisis of the 1970s, Germany's policy approach addressed research into alternatives to conventional energy, and the 1986 Chernobyl nuclear accident cemented the disenchantment with nuclear power. In March 1987 Chancellor Kohl declared that the climate issue represented the most important environmental problem. On the national level the Committee for the Environment, Nature Conservation, and Nuclear Safety of the German Bundestag established an Enquete Commission on Preventive Measures to Protect the Earth's Atmosphere, with the mandate to study the ozone problem as well as climate change and to make proposals for action. An inter-ministerial working group on CO_2 reduction was also established.

Targets for reductions of GHGs were set, and a series of proposals, which included an electricity feed-in law for generation from RES, were formulated. There was a growing consensus among MPs of all party groups that it was time to create markets for renewable energy technologies. The measures consisted of the 100/250MW wind programme, the '1000 Solar Roof programme' and the establishment of a legal basis for utilities to pay higher prices for RE.

From 1991 to 1995, under the 1000 Solar Roof programme, applicants received 50 per cent of investment costs from the federal government plus 20 per cent from the regional government. Eventually, 2250 roofs were equipped with PV modules, leading to about 5MW of installed capacity. As to wind energy, a programme for subsidizing 100MW – later 250MW – wind turbines (with a rate of €0.04/kWh, later reduced to €0.03) was necessary to gain practical experience with different approaches under real life conditions. As this programme in 1991 combined with the Feed-in Law, installed wind capacity grew rapidly. In subsequent years, these subsidies declined rapidly (Mez, 2004).

The red–green coalition government decided in 2001 on a plan to phase out the operation of German nuclear plants, providing further impetus to the renewables industry.

CREATION OF THE FIRST FEED-IN LAW

The first proto-feed-in law was actually created in the US and dramatically stimulated the wind industry in some states, especially California. PURPA, or Public Utility Regulatory Policies Act, was a US federal law enacted in 1978 that was intended to encourage more energy-efficient and environmentally friendly commercial energy production. PURPA defined a new class of energy producer, called a qualifying facility. When a facility of this type meets the Federal Energy Regulatory Commission's requirements for ownership, size and efficiency, utility companies are obliged to purchase energy from these facilities based on a pricing structure referred to as 'avoided cost' rates. Avoided cost is the marginal cost for the same amount of energy acquired through another means, such as construction of a new production facility or purchase from an alternative supplier. For example, a megawatt-hour's (MWh's) avoided cost is the relative amount it would cost a customer to acquire this energy through the development of a new generating facility or acquisition of a new supplier. These rates are usually quite generous for the producer.

Since 1979 Germany had made efforts to stimulate demand for renewable energy by use of a tariff system, not dissimilar to the US law. At that time the German government relied on the national competition law to oblige electricity distributors to purchase electricity from renewable sources produced in their area of supply, again based on the principle of avoided costs.

Historically, the conflict between political goals and the business interests of the energy providers in Germany would usually be solved by free market competition. However, that is only partially possible because of the monopolistic position of a grid-based supply structure, in that only one owner can control any one section of the grid. Any possible competition was prevented by political regulations laid down in the Energy Industry Law of 1935. Furthermore, the feed-in of electricity into the grid of an electricity provider was not regulated by the law, but was left to contractual negotiations between the local electricity provider and the supplier.

The so-called Association Agreement of 1979, amended in 1985, was meant to fill the gaps regarding the feed-in of industrially produced electricity into the grid of the electricity providers. This agreement mainly considers the interests of large industrial enterprises with their own large power stations (for example the steel or chemical industry), but not decentralized generation in smaller hydroelectric power plants, wind turbines, biomass plants or communal heating/power stations.

Under these circumstances, sufficiently profitable operation was prevented from the outset, even if under free market competition it could have been profitable. Also, most of the supervisory federal and Länder (regional) ministries (which were traditionally closely tied with the interests of the energy industry) did not want to facilitate the economic interests of the new electricity providers. The energy industry viewed decentralized electricity generation as competition and acted accordingly.

In the Länder ministries responsible for energy policy, the arguments cumulated especially within the working group energy policy of the Ministry of Trade and Industry. Therefore it became a matter of discussion for the ministers. However, after nearly ten years of discussion there was still no breakthrough, particularly because the leading federal ministry of trade and industry largely adopted the position of the electricity providers. With that background it was reasonable but also surprising that from the centre of the German Parliament a cross-party initiative to create a feed-in-law was started (Kissel and Oeliger, 2004).

THE 1990 FEED-IN LAW: STROMEINSPEISUNGSGESETZ

The first feed-in law in Germany took the form of a relatively simple one-page bill for assisting producers of electricity from small hydro stations, primarily in southern Germany. The nascent wind energy industry got involved in the initiative, so that the law could be adopted for small hydro and wind energy installations. During this period many discussions were held to convince the sceptics that RE from small decentralized generators could be re-introduced or newly built only through legal regulation.

The law met resistance from the Federal Ministry of Economics and Technology, but in parliament the idea gained acceptance; support also came from the Ministries of Research and of the Environment. The bill gained consent from all parliamentary parties and became the Electricity Feed-in Law of 1990, or Stromeinspeisungsgesetz (StrEG).

There was no initial reaction from the large utility companies, perhaps because they sensed no threat to their position from such a limited piece of legislation. The law required electric utilities to connect RES generators to the grid and to buy the electricity at rates of 65 to 90 per cent of the average tariff for final customers. Generators were not required to negotiate contracts or otherwise engage in much bureaucratic activity.

Together with the 100/250MW programme and subsidies from various state programmes, the law gave considerable financial incentives to investors, although less for solar power due to the high cost. One of the stated purposes of the law was to 'level the playing field' for RES by setting FITs that took account of the external costs of conventional power generation. In parliament external costs of about 3–5 cents per kWh for coal-based electricity were mentioned by MPs from the Christian Democratic Union (CDU) party. Before adoption, the law was notified to the European Commission for approval under state aid provisions. The Commission decided not to raise any objections because its effects were insignificant, and because it was in line with the policy objectives of the Community. However, it announced that it would examine the law after two years of operation.

The StrEG was a good starting point for a fair price for electricity from RE. The minimum reimbursement per kWh for electricity from solar and wind energy amounted to at least 90 per cent of the average price electricity providers charged the end consumer. This, however, only allowed profitable business at very good locations for wind power operators. Inland wind plants and especially PV plants could not be operated profitably under the conditions established within the StrEG. Against this background many cities and parishes introduced additional reimbursements to allow an economically viable operation as incentive for investments.

At the beginning of 1990 the production of solar electricity was at about ten times the costs of coal or nuclear electricity despite the StrEG, and therefore did not stand a chance in the free market. The new law was one of the first successful support mechanisms for RE technologies. With that regulation in place, the operators of solar and wind installations could balance their entire costs, including expenses for financing, with the revenue from production. The minimum reimbursement of approximately 17 pfennig per kWh fed, which then was paid by the electricity providers, was clearly below the supply cost of solar electricity and was therefore raised to 2DM in the law. Therefore the lengthy procedure for promoting solar energy with subsidies became unnecessary.

In 1989 a change to the federal standard rates for electricity by the upper house of parliament made an equitable reimbursement possible for progressive electricity providers on a voluntary basis. This payment had the disadvantage of needing the goodwill of electricity providers to accommodate RE generation. This goodwill was, however, missing for a long time, leaving many solar pioneers empty-handed. In addition, the StrEG did not include guaranteed tariffs for all renewable sources, for example biomass.

Solar PV was not assisted by the new law in the same way as wind. Wind benefited from a combination of the law and a 250MW promotion programme. Solar had a subsidy programme for 1000 Solar Roofs, but this failed to achieve a breakthrough for the technology because the tariff levels were set too low, and a further programme was not initially forthcoming. This created a poor investment environment for solar, and market volume remained low.

Then, however, a 1989 modification of the federal framework regulation on electricity tariffs permitted utilities to include cost-covering contracts for electricity using RE technologies, even if these 'full cost rates' exceeded the long-term avoided costs of the utilities concerned.

Further impetus to the solar market was provided by a combination of local activism, regional government action and Greenpeace initiatives. All of this activity resulted in stabilization and growth of the solar market (Kissel and Oeliger, 2004).

COMPLEMENTARY SUPPORT MEASURES

A federal energy research programme from 1990 to 1998 provided more than €1 billion for all RE technologies. The Länder contributed another €0.85 billion for the period 1990–1997. Loan programmes by the federal government's banking institutions Deutsche Ausgleichsbank and Kreditanstalt für Wiederaufbau permitted more than €3 billion in reduced interest loans for RES installations in the period 1990–1998. Other measures privileged wind turbines under the construction code (every local community had to present a plan with zones appropriate for wind power, which greatly facilitated permitting), reformed training programmes for architects, and developed public awareness campaigns. The reforms to the federal building codes were of major significance. The changes prioritized the generation of electricity from installations outside of buildings. This law has been very effective in overcoming protests against wind farms.

THE StrEG's MAIN ISSUES OF CONTENTION

These incentives, and the market growth they spurred, encouraged technological and political learning in this sector, but also strengthened the resolve of the large supra-regional utilities to attempt a rollback of this law, via both politics and the judiciary. This was more than just opposition to small and decentralized generation. Firstly, no provision had been made to spread the burden of the law evenly in geographical terms; a satisfactory solution to this problem did not materialize until 2000. Secondly, the utilities were by this time marked by the experience of subsidies for hard coal used in electricity generation, which had grown from €0.4 billion in 1975 to more than €4 billion annually in the early 1990s. Two thirds of this was covered by a special levy on electricity, and one third had to be paid by the utilities directly but was also passed on to the consumers (Lauber and Mez, 2004, p4).

In April 1998 the Energy Supply Industry Act was adopted to transpose the European electricity directive 96/92/EC and modified the StrEG in several ways. In particular, it created a new compensation mechanism for distributing the supplementary cost to the utilities. The 1990 law had included a hardship clause which was never practically applied. Wherever RES exceeded 5 per cent ('first ceiling') of the total electricity supply, the upstream network operator had to compensate that undertaking for the supplementary costs caused by this excess amount.

A similar rule applied in favour of the upstream network operator, who could request compensation from a network operator situated further upstream if the compensation he had to pay exceeded 5 per cent of his output ('second ceiling'). As it was obvious that in some coastal areas the 10 per cent limit would be reached, wind power growth could stop unless an alternative solution was found. This conflict led to insecurity for investors and stagnating markets for wind turbines from 1996 to 1998.

In the past there have been major, sustained protests against the StrEG. The big energy companies twice succeeded in bringing their protests before the Bundesverfassungsgericht (Federal Constitutional Court), and once approached the European Court. Their argument was that privileging RE would be opposed to the principle of market equality. The court stated that privileging renewable energies should be allowed due to their benefits to the environment.

THE 2000 RENEWABLE ENERGY SOURCES ACT: ERNEUERBARE-ENERGIEN-GESETZ

In response to deregulation of the electricity market in 1998, and in view of a number of problems with the StrEG, German RE policy was updated, refined and replaced by the Erneuerbare-Energien-Gesetz (EEG), or Renewable Energy Law, in 2000. The aim of the EEG reads: 'Enabling in the interest of climate

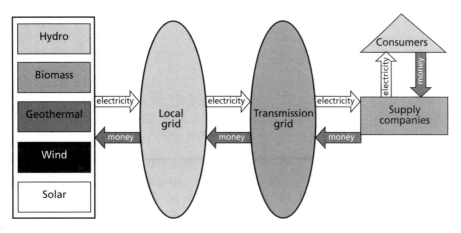

Source: Viertl, 2005

Figure 4.1 *How the EEG works*

and environmental protection, the development of a sustainable energy supply to increase the contribution of renewable energies to the electricity supply, according to the goals of the European Union and the Federal Republic of Germany to at least double the contribution of renewable energy to total energy consumption by 2010' (§1, EEG, 2000).

Figure 4.1 shows the simple functioning of the system. Electricity is generated, distributed and consumed, flowing one way; reimbursement flows in the other direction.

In summary, the key changes were in the differentiation of the tariffs, depending on RE type, size and site, and the replacement of percentage-based tariffs with fixed rates over fixed periods. Because the payment to RE producers was a percentage of the final consumer price under the StrEG, when deregulation brought prices down, it simultaneously reduced incomes for RE producers.

The first amendment of the EEG extended the range of technologies to be covered, to include geothermal energy and mine gas. The power limit for hydro plants and installations using sewage or landfill gas of 5MW fixed in the StrEG now also applied to installations based on mine gas or solar energy. In contrast, the power limit for biomass plants was raised from 5 to 20MW, allowing larger plants to benefit.

To stimulate technological innovation and efficiency, and to ensure better compatibility with the European law on state aid, the reimbursement period was limited and the remuneration paid under the EEG also included a digressive element: From 2002 on, new installations of biomass (minus 1 per cent), wind (minus 1.5 per cent) and PV (minus 5 per cent) received lower tariffs. From 2003 on, new installations of these types received tariffs lowered by a further 1, 1.5 or 5 per cent, and so on for the following years.

The most far-reaching changes, in comparison to the StrEG, were related to the remuneration scheme. While the StrEG was still based on uniform minimum reimbursement, the production costs of RE sources became the relevant criteria for the EEG tariffs (Kissel and Oeliger, 2004). Scientific studies determined the figure that would allow profitable operation with state of the art technology and geographical advantage. The EEG raised all remuneration rates, but differentiated them according to the source of energy, capacity or location of the plant. Except for hydropower, where the amortization of the power plants normally takes several decades, the EEG fixed the purchase guarantee and the FITs for 20 years from the start of operation of every new qualifying RE plant. Due to the high production costs of PV electricity, the operators were paid the highest reimbursement, set at 99 pfennig per kWh fed. That amount was achieved by the 100,000 roof installation programme, which was included in the 1998 government programme. The programme started on 1 January 1999 with preferential credits starting at zero interest and the foregoing of 12.5 per cent repayment of the credit sum. The programme did not take off without higher reimbursement rates. In order to ensure success, complementary reimbursement rates could be enforced – at the time limited to the size of the 100,000 roof installation programme. The programme was subject to an annual 5 per cent degression, as due to production optimization considerable cost degressions were expected with new installations.

The EEG gave all operators with the exception of public authorities the opportunity to claim from the EEG. This enabled the municipal companies, as an important player within the electricity sector, to participate as investors in the development of electricity generation from renewable sources.

As a consequence of disproportionate generation of wind power in the north of Germany, the electricity providers and the consumers in that area were loaded more than others with the additional costs. To a lesser extent that was also valid in southern Germany due to the above average usage of hydropower. In considering the EEG, the federal parliament considered that a proportional cap by supply area (as introduced in the second amendment of the StrEG in 1998) was insufficient, 'because the use of wind power already exceeded the limits of a market launch in North Germany' (EEG: Explanatory Statement of the Federal Parliament – General Section, in: SZ 1/100, p9). In §11 of the EEG a country-wide compensation scheme was incorporated to help balance the costs between the different electricity providers. This meant that all electricity providers reached the same proportional balance between the EEG electricity and total electricity through compensation, with none carrying extra costs, even if their region produced more electricity from renewables.

To comply with European law, the EEG set three further provisions. Firstly, by 30 June every two years after the entry into force of the law, a report shall be submitted on the progress achieved in terms of the market introduction and the cost development of RES power generation plants. Where necessary, this report shall propose adjustments of the remuneration amounts and of their reduction rates, in keeping with technological progress and market developments with regard to new installations.

Secondly, relating to the remuneration for wind power, the different quality of plant sites was taken into account ('Referenzertragsmodell'). The purpose of this new provision is to avoid payment of compensation rates that are higher than what is required for a cost-effective operation of such installations, and to create an incentive for installing wind energy converters at inland sites. Operators of onshore wind turbines receive a fixed FIT during the first five years after the plant has started operating. The EEG defines a reference wind turbine, which is located at a site with a wind speed of 5.5 metres per second at an altitude of 30 metres. This reference turbine would generate a so-called reference yield in a five-year period. If a wind turbine produces at least 150 per cent of this reference yield within the first five years of operation, the tariff level will be reduced for the remaining 15 years of support. However, for each 0.75 per cent the generated electricity stays below the reference yield, the higher starting tariff will be paid for two further months. This means that the use of wind energy to generate electricity is not restricted to locations with very good wind conditions but that sites with less favourable conditions can also be exploited. Offshore plants, starting to operate before the end of 2010, receive a higher starting remuneration for the first 12 years. This period will be extended, if the wind turbine is positioned more than 12 nautical miles away from the coastline and if the water is more than 20 metres deep. For each mile the distance to the coast line exceeds 12 miles, the period of higher remuneration will be extended by 0.5 months. For every metre of water depth that exceeds 20 metres, it will be extended by 1.7 months. Consequently higher expenses for constructing wind turbines at a greater distance from the coast line or in deeper water and for their connection to the electricity grid are taken into account (Klein et al, 2006).

Thirdly, the remuneration scheme for PV power contained a special provision that is connected with the compliance with the European law. The guaranteed remuneration shall not apply to PV systems commissioned after 31 December of the year following the year in which PV systems reach a total installed capacity of 350MW. This limit was raised to 1000MW in June 2002 because the 350MW seemed to be surpassed already in 2003 and the successful PV sector needed further planning security.

The Act is very effective, because the costs for RE hinge largely on investment security. If an investment is high risk, banks demand high interest rates for the loan and the investors demand high-risk mark-ups. Since the structure of the EEG guarantees a particularly high investment security, credit interest rates and risk mark-ups are low compared with other instruments. Furthermore, the lowering of fees as laid down in the EEG for installations commissioned at a later date ensures further price reductions. This degression has already had an impact: The costs for installing PV systems dropped by 25 per cent between 1999 and 2004; for wind turbines, costs were reduced by 30 per cent between

1993 and 2003. The degression also leads to installations being constructed as quickly as possible, in order to obtain a high payment level, and deters operators from waiting until installations become cheaper. The EEG ensures very high-quality installations as – because payment is made per kWh produced – there is great incentive for operators to run their installations efficiently and with as little interruption of operation as possible, certainly during the initial 20-year payment period. Operators therefore demand high standards from the installation manufacturers (Federal Ministry for the Environment, Nature Conservation and Nuclear Safety (BMU), 2004).

THE 2004 EEG AMENDMENT

The amendments to the EEG display clearly the national commitment to renewables. The stated aims of the amended EEG are to 'increase the share of renewable energies in the total electricity supply to at least 12.5 per cent by the year 2010 and to at least 20 per cent by the year 2020, and the further development of technologies for the generation of electricity from renewable energies, thus contributing to the reduction in costs' (BMU, 2004). The EEG amendment also helps transpose the September 2001 European Union directive on the promotion of RE in the electricity sector, ensuring that all RES defined in the directive are covered by the EEG.

After the re-election of the red–green coalition in autumn 2002, responsibility for RES changed from the Economic Affairs Ministry (held by a Social Democrat) to the Environment Ministry (held by a Green); the parliamentary committee in charge changed in a parallel fashion, presenting new opportunities. The first draft of the amendment by the Environment Ministry led to a lively conflict with the Economic Affairs minister, a well-versed politician from the coal state of North Rhine-Westphalia.

The minister attacked the very principle of the FIT, recommending its replacement with a tender system, and arguing that rates were excessive, especially for wind energy. His main concern seems to have been to protect the vested interests of large conventional energy companies. After a compromise within the government, the red–green majority in parliament proceeded to revise the government bill largely against the preferences of the minister. However, he was successful in obtaining reduced rates for wind. In the Bundesrat, the Länder ruled by conservative governments opposed the bill. The Bundestag majority could simply have insisted on its earlier version. However, the red–green coalition negotiated with the conservatives in an effort to secure support for maintaining the EEG beyond 2007. Some of the Länder wanted an expiration date of 2007 for the Act, or a declaration reversing the nuclear energy phase-out; some criticized the 20 per cent RES target for 2020. Finally, however, the Conciliation Committee was content with more modest changes, and the bill was adopted in both houses (Mez, 2004).

In the amendment, a clear regulation concerning grid costs was established. The costs for grid connection must be paid by the plant operators, and costs for upgrading the grid must be borne by the grid operator. For the settlement of any dispute in relation to grid costs, the Federal Ministry of Economics and Technology (BMWi) established a clearing centre, with the involvement of the parties concerned.

The equalization provision in the EEG is aimed at the operators of transmission grids. (This is a small group with a limited number of players who will easily be able to handle the transactions associated with the equalization scheme, and will also be able to monitor each other.)

Compared with the previous EEG, the amendment provides for a more differentiated fee structure, taking account of efficiency aspects. In particular, the payment conditions for biomass, biogas, geothermal and PV energy were improved. If existing large hydropower plants are modernized or expanded (up to a capacity of 150MW), the additional electricity generated is included in the fee.

For 2005, fees under the amended EEG range from 5.39 €cents/kWh for electricity from wind energy (basic payment) to 59.53 €cents/kWh for solar electricity from small facade systems. The annual degression in the fees for new installations was increased to strengthen the incentives for technical innovations and cost cutting – for example, 2 per cent for wind energy, 1.5 per cent for bioenergy and 5 per cent for PV energy, starting from 2005. From 2006 onwards, the degression for new PV installations on open spaces was even increased up to 6.5 per cent. For the first time also a degression of 1 per cent for new geothermal plants was introduced, starting from 2010.

For the area of bioenergy, in addition to the minimum fees laid down, the amendment provides for additional fees (bonuses), if the electricity is exclusively produced from self-regenerating raw materials, combined heat and power (CHP), or if the biomass was converted using innovative technologies (e.g. thermal chemical gasification, fuel cells, gas turbines, organic Rankine systems, Kalena cycle plants or Stirling engines). The bonuses can be used cumulatively. The payment rate for wind energy on land was lowered in the amendment. Wind parks that could not achieve at least 60 per cent of the reference (expected) yield at the planned location can no longer claim payment under the 2004 law.

For coastal sites in particular, there are new incentives for 'repowering' (the replacement of old, smaller installations by modern, more efficient ones). The starting fees for offshore wind parks are paid for installations commissioned prior to 2011. The period for the higher starting fee for offshore wind parks is a minimum of 12 years. This period can also be extended for installations located further from the coastline and erected in deeper water (BMU, 2004, p5).

The increase in tariffs paid for solar PV was perhaps the most important of the changes, making PV far more attractive commercially and leading to a solar boom in 2004.

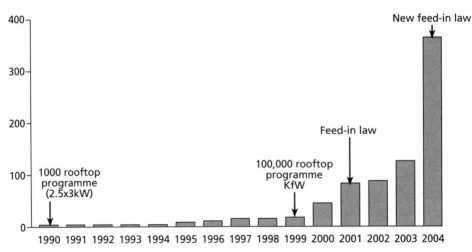

Figure 4.2 *Effects of the German market launch programmes on RE capacity growth*

Since July 2003 there has been an equalization regulation for electricity-intensive companies in the producing sector. This regulation was expanded in the amended EEG. Electricity-intensive companies in the producing sector and environmentally friendly railways can be included under the equalization regulation if their electricity consumption is higher than 10GW (previously 100GW), and the ratio of electricity costs to gross value added exceeds 15 per cent (previously 20). The amendment to the EEG limits the total relief volume. This again limits the extra costs incurred by non-privileged companies due to the equalization scheme. The electricity volumes that are distributed among the non-privileged electricity consumers are limited to a maximum of 10 per cent above the share calculated pursuant to the EEG.

Figure 4.2 shows the effects of the German legislation on PV deployment. Note also the effects of combining support measures.

THE NEW ENERGY POLICY OF THE RED–GREEN COALITION

It is worth highlighting briefly the raft of legislative measures and incentive programmes that the red–green coalition introduced or worked within to support the expansion of renewables in the national energy mix. Single policies are rarely able to achieve everything that proponents would like, and often have to be supplemented with other discrete programmes. This draws again on Mez (2004).

This coalition government emphasized ecological modernization, climate change policy, job creation and socio-economic development; energy policy was to be a leading example. It included tax reform (introducing an eco-tax on energy), phasing out nuclear power and strengthening of RES and CHP. Additional reform of the Energy Supply Act and of the Association Agreements followed in a second phase. This led the government to agree to the obligatory provision of a regulator in the new electricity directive of 2003, to be implemented in 2004.

Nuclear power phase-out

The 2001 Nuclear Energy Phase-Out Act effectively ended new build, and licences of existing plants were reviewed and limited in time. The legislative process was characterized by the government's endeavour to reach a consensus with nuclear power interests and to avoid legal disputes before the courts.

Climate change policy

Within the framework of the Kyoto Protocol and the European burden-sharing agreement, Germany committed to reduce GHG emissions by 21 per cent from 1990 to 2008/12. In addition, the government in 1995 had pledged a 25 per cent reduction of CO_2 emissions by 2005. By 2000, a reduction of about 18 to 20 per cent, corresponding to 180 to 200 million tonnes of CO_2, was already achieved, so that the gap amounted to 50 to 70 million tonnes of additional reduction. This was to be achieved by the government's Climate Change Policy Action Programme of October 2000. Both the EEG and the CHP Act are integral parts of this programme. These two areas of activity are expected to contribute reductions of 15 million tonnes and 23 million tonnes of CO_2 respectively, or about 50 per cent of the target (BMU 2000, pp9, 77, 80).

The eco-tax reform

This reform was passed as one of the first environmental initiatives of the new government in two consecutive laws which introduced a tax on the consumption of electricity (at a reduced rate for industry) and raised existing mineral oil taxes (i.e. on petrol, diesel, natural gas and various mineral oils). Tax levels for petrol and diesel, as well as electricity, increased in five steps up to 2003. Coal and nuclear fuels were not affected. The tax is not levied on fuels used in CHP and decentralized production (up to 5MW), nor for natural gas-fuelled power plants with an efficiency of 57.5 per cent or more. The advantage for these sources is up to 1.53 €cents/kWh. But on the, at times, low price market, this was not sufficient to bring about their expansion.

The bulk of the revenue – rising from €4.3 billion in 1999 (€8.8 billion in 2000, €11.8 billion in 2001 and €14.3 billion in 2002) to €18.7 billion in 2003 – is earmarked to lower the retirement pension contributions from employees as well as employers, thereby lowering the production factor cost of labour while increasing that of energy. A small amount of about €102 million per year (1999 and 2000) was reserved for RE subsidies, particularly to finance the 100,000 roof programme. The promotion of RES increased to €153 million in 2001, €190 million in 2002 and €250 million in 2003. The eco-tax reform is expected to reduce GHG reductions by about 2 to 3 per cent by 2005. For 2002, its impact on CO_2 reduction stood at 7 million tonnes.

Combined heat and power and end-use efficiency

The efforts to increase efficiency are also reflected in support for CHP, the share of which is to increase from 12 per cent in 1999, substantially below that of other European countries. Denmark saw strong growth in biomass until 2001 with large centralized CHP plants, initiated by the relatively high FITs and a stable policy framework (European Commission, 2005). CHP plants are under severe pressure since electricity liberalization. The new act for the support of CHP plants for public supply entered into force in April 2002 and was designed to create incentives for modernization until 2010, leading to a reduction of some 11 million tonnes of CO_2. Additional support is provided for small-scale CHP and fuel cells. As to end-use efficiency, activities were initiated in line with EU policy. As a first step, the Energy Savings Ordinance entered into force in February 2002. It set the total energy requirement of new buildings at 30 per cent below current standards; for old buildings, insulation requirements and exchange of heating systems were prescribed.

Renewable energy targets

The government formulated a target to increase the share of RES in the electricity supply to 12.5 per cent in 2010 and 50 per cent in 2050; in 2004 the goal of 20 per cent by 2020 was added. The long-term target must be viewed as a programmatic goal, which in concert with energy efficiency programmes is ambitious but not unrealistic either technically or economically. Several measures were taken in favour of RE. They included a five-year market incentive programme for RES which provided about €445 million from 1999 to 2002. A tax break on biofuels was applied in keeping with an EU directive on the subject. On the international level, the German government in 2004 hosted the international conference on RE in Bonn. As to RES, the most important measures adopted were the 100,000 roof programme for PV and the EEG for all technologies.

The 100,000 roof programme

Solar PV had not been able to develop significantly during the 1990s. The red–green government wished to provide new impetus. As the design of a new feed-in regulation was expected to take time, another market creation programme along the lines of the 100MW wind and 1000 roof programme (both 1989) was adopted in January 1999 as a stop-gap measure. It provided for improved loans for PV roof installations; the goal was to achieve an installed capacity of about 300MW. The programme was taken up slowly at first, but took off when the EEG was introduced. By 2003, the two measures had led to installations of 350MW. At that point, the 100,000 roof programme was terminated and PV market development was then supported by improved FITs.

The EEG clearly recognizes the contribution of RE to reducing GHG emissions and saving scarce fossil fuel reserves. It aims to initiate a self-sustaining market for RE by compensation for the distortions in the conventional electricity market, and simultaneously creating a critical mass by means of a massive market introduction programme that does not lead to any additional burden on the taxpayer. This price regulation, in combination with measures for the internalization of external costs (e.g. the eco-tax reform), is designed, over the long and medium term, to ensure competitiveness with conventional energy sources. Thus, in addition to environmental and climate-policy targets, the law clearly pursues industrial policy goals (REACT, 2004, p2).

Criticism of the EEG

At its June 2005 congress in Berlin, the German Electricity Association (VDEW) introduced a 'Discussion proposal on the future support of renewable energies'. According to the proposal, the Renewable Energy Sources Act (EEG) would be abolished and replaced by another (quota) instrument. On 6 October, at the European Conference on Green Power Marketing, Roger Kohlmann, Deputy CEO of the VDEW, tried to back up the supposed advantages of the VDEW proposal scientifically and prove them with concrete numbers. The basis for his arguments was an expert opinion provided by B. Hillebrand (EEFA – Energy Environment Forecast Analysis), which had been commissioned by the VDEW ('The integrative model for renewable energy sources – from state to market').

The BMU issued a paper in response, turning the arguments on their heads, and demonstrating that the proposals would in fact be counter-productive for the German renewables industry and result in greater expense for end-consumers. The following comments from the BMU on the VDEW proposals concern the arguments levelled at the EEG:

- *The EEG is too expensive. The so-called VDEW 'integrative model' could reduce costs significantly.*
  On the basis of the expert opinion it commissioned, the VDEW claims that the 'integrative model' could save a total of 3.8 billion euro by 2020. That would be an average of about €270 million every year. In addition, it suggests economic efficiency gains of up to €1.7 billion due to decreasing

prices for electricity. However, numerous studies and publications claim the exact opposite. As long as the principle of expansion is adhered to, including the expansion of individual branches of renewable energies, switching systems in favour of a quota system – which is the aim of the 'integrative model' – would lead to higher costs but lower effectiveness. The German Institute for Economic Research stated in its 29th Weekly Report 2005: 'In case of a pre-determined target for the expansion of renewable energies the financial burden for the electricity consumer can be controlled especially by means of differentiated and degressive rates of support. Between 2000 and 2004 these costs were lower by a total of 1.7 billion euro than they would have been in the case of a uniform support such as a national quota.' This means that within this time span electricity consumers saved an average of 340 million euro with the EEG compared to the integrative model.

Under the so-called 'integrative model' the savings cited above by the VDEW would be achieved in particular by stopping the expansion in the solar energy and biomass branches (by using specifically cultivated regenerative raw materials). It should be borne in mind that according to a survey carried out by the Forsa Society in April 2005, 62 per cent of Germans want a stronger expansion of renewable energies and 25 per cent agree with current policies. 85 per cent feel that in 20–30 years the energy supply should be provided by solar energy. Furthermore, against the background of falling prices for agricultural products on a globalized market German farmers are glad to be able to develop a second source of income as energy farmers.

While stopping the expansion of photovoltaic and biomass plants would therefore be backed neither by a majority of the population nor by a majority in the Bundestag, steering in that direction could be achieved without completely changing the system. The current version of the EEG provides for enough possibilities to control the speed of expansion of individual branches. The advantages the EEG has for the national economy would be maintained. For the above technology the 'integrative model' would be more expensive than the EEG (see above, www.erneuerbareenergien.de/inhalt/36040/4595/).

- *The EEG contravenes market rules.*
Critics reject the EEG as a contravention of market rules. In fact all instruments for the market introduction of renewable energies directly intervene in the market. The EEG lays down fees, quota systems control the amount, thereby steering the market. Quota schemes oblige the energy supplier to provide a certain share of the electricity through renewable energies – or to buy expensive certificates instead. In contrast to countries which apply other schemes, the growth in the renewable energy sector triggered by the EEG has created a large market among plant manufacturers and suppliers in Germany. The competition created and the experience gained by increased sales numbers have led to decreasing prices. The degression laid

down in the EEG – every year the minimum prices for newly commissioned plants decrease by 1–6.5 per cent – fuel the competition and therefore the dynamics of cost reduction even more.

- *The EEG leads to an additional need for regulation energy and the need for a vast extension of the grid.*
  According to a fundamental study by the German Energy Agency (dena grid study) which was drawn up jointly with the public utilities, representatives from the renewable energies sector and the competent federal ministries, various sections of grid, adding up to about 400km, will have to be upgraded by 2020 at the latest. The construction of about 850km of completely new lines will be necessary. Other additional measures to improve the grid are necessary. The existing extra high voltage grid will have to be extended by about 5 per cent.

  The study assesses the extension of the extra high voltage grid necessary for wind energy as technologically and economically feasible. The study further states that the level of the additional flexibility needed can be handled. The need for regulation and reserves caused by the expansion of wind energy can be provided by the power plant park further developed according to the dena grid study.

  Furthermore, the issue of whether and how much grid capacity is needed generally is independent of the instrument used for the expansion of renewable energies. For example, if under a quota system there was a great increase in wind energy capacities mainly in the north of Germany where there are strong winds, this electricity would have to be transported to the consumer – thus requiring an extension of the grid.

  The differentiation of fees provided for in the EEG enables plants to be operated not only in sites with strong wind, thus allowing an electricity production closer to the consumers. In addition, the EEG allows electricity from renewable energy sources to be produced close to the consumer everywhere in Germany by means of photovoltaics as well as the branches of biomass and geothermal energy which are able to provide a base load. Since the integrative model would significantly reduce the expansion of these three branches, it would aggravate the problems concerning grid capacities and regulation energy.

- *The expansion of renewable energies allows only few conventional power plants to be decommissioned.*
  Concerning the expansion of wind power this statement is in principle correct. However the reduction of greenhouse gas emissions and the goal to become more independent of fossil energy sources and their fluctuating or rising prices decisively depends on whether the use of fossil sources can be reduced and thus CO_2 emissions avoided. And this is the case: the EEG saved about 33 million tonnes of CO_2 in 2004. By 2010 it will be at least 50 million tonnes. The only other instrument which achieves reductions at a similar level in Germany is the ecological tax reform.

- *The EEG does not fit into the European internal market from a legal point of view.* The VDEW claims that within the framework of the European internal market it is imperative for legal reasons to harmonize the instruments used by the different European Member States. This claim is wrong, as Prof. Dr. Stefan Klinski expressly confirmed in his legal opinion (http://www.erneuerbare energien.de/inhalt/36005/4596/). Furthermore the European Court of Justice expressly confirmed the admissibility of the EEG in a decision of 2001. On the contrary, the EU Directive on the promotion of electricity produced from renewable energy sources has decided on coexistence and competition between the different national instruments. In this competition the EEG and the feed-in tariff system are way ahead of any quota system which the VDEW wants to introduce as a uniform system on the European level: most EU Member States use a feed-in system because it is considerably less expensive than existing quota systems and in contrast to these actually leads to an expansion of renewable energies. According to statements by EU Energy Commissioner Andris Piebalgs, the Commission will therefore not propose a harmonization in the near future. At most a coordination of the different instruments of the European Member States is under discussion.

  With regard to the community law's principle of subsidiarity there are no economic or legal arguments against the use of differently designed instruments for the promotion of renewable energies in the single Member States on a future deregulated European energy market.

The aim of the European Community is to expand RE equally everywhere in Europe. Therefore, the EU Directive on the promotion of electricity produced from RES deliberately lays down a national indicative target for every Member State which is oriented around the respective use of RE and the usable potential in each country. The European Commission accordingly records and assesses the production of electricity from RES in each country. There are good reasons for an expansion of RE in each country:

- In the long term the share of RE on a European level should exceed the 21 per cent planned for 2010. In order to make this possible the existing potential in all countries must be exploited.
- Concentrating the expansion of RE in single Member States might jeopardize society's acceptance of RE and would contradict the common responsibility of all Member States for effective climate protection. Apart from this, the transport of electricity over long distances does not make sense either from an economic or an ecological point of view.
- In the countries practising it, the expansion of RE creates jobs and turnover in the respective economic branches (BMU, 2005).

EFFECTS OF THE LEGISLATION

The German feed-in laws have consistently added to job growth, even during times of rising national unemployment; they have also produced valuable export markets, increased energy security and, most crucially of all, have prevented the emission of vast quantities of GHGs. German energy use has been comparatively stable, and so the increase in use of RE has had a directly beneficial effect on the level of global emissions.

Keeping up to date with the positive effects of RE use in Germany is challenging due to their rapid growth rate, but recent figures demonstrate the ongoing success of the market launch programmes in the country, and the commitment of many actors to making Germany a more sustainable, secure and safe nation. Making an entire nation *completely* sustainable is a challenge of a greater magnitude.

To put this in some perspective, it is worth restating that we are entering a time of terminal decline in oil production, while global demand for energy and all resources is booming, due largely to the explosion of economic development in China, other South East Asia countries and India. The International Energy Agency forecasts that annual demand for energy will rise by 59 per cent by 2030, yet known oil and uranium reserves will last only around 50 years each, and gas reserves only 66 years. Many nations and individuals are already in 'fuel poverty', spending up to 10 per cent of their income on fuel. Under a market system, rocketing demand and plummeting availability can only mean one thing: conventional energy prices will continue to rise, consuming an increasing proportion of spending at all levels. RE technology by contrast, is improving fast, with technological breakthroughs occurring almost monthly. Costs are therefore falling, and ever greater efficiency is being achieved. Making the shift from fossil and nuclear to renewables sooner rather than later is essential in not only forestalling climate chaos, but also in improving the fortunes of individual nations and their citizens.

Germany and Spain, as world leaders in RE production, have achieved great things through their FIT systems, and almost all old installations in Denmark are based on this system. Denmark is currently the only EU country that is a net exporter of energy, and has kept consumption flat over the last 25 years even as its economy has grown by 50 per cent. Low wastage and big investment in RES such as wind farms have contributed to that success. Nearly 78 per cent of all wind power capacity in the European Union, amounting to 40,504MW at the end of 2005, was installed in these three countries (Bechberger and Reiche, 2006).

The solar sector in Germany has grown substantially. With its 47 per cent share, Germany is the largest solar heating market, with highly developed systems. In addition to the small- to medium-sized producers of solar power systems, those also providing solar power systems are the heating industry and construction companies such as roof-tile and roof window manufacturers and facade companies. A growing number of component manufacturers supply globally. Use of solar heat is being promoted by a market-stimulation programme of the German environment ministry. For plants that heat drinking water, investors obtain a subsidy of €135 per square metre of collector area. This is about 15 per cent of the investment outlay. The solar heat sector employs about 10,000 staff throughout Germany in the production of components and plant, marketing and installation of solar heating systems, achieving a turnover of €550 million in 2004.

Since the start of the 100,000 roof scheme in 1999, and of the EEG coming into force in 2000, the market for solar power has experienced vigorous growth in Germany. In 2003, the 100,000 roof scheme terminated. Since 2004, they are promoted only with increased compensation through the EEG, with no more subsidies from the federal budget. Through the government's market-launch policy, a modern and forward-looking industry was able to develop. Over 40 companies produce solar power components at all stages of the production chain. A number of companies are listed on the stock exchange and are among the top companies worldwide. The solar power sector employs 20,000 people in Germany, achieving a turnover of €1.7 billion in 2004. Between 2000 and 2005, the turnover of the German solar power sector increased by an average of 43 per cent per annum. A turnover of €2.7 billion was anticipated for 2005 (Friedrich Ebert Foundation, 2006).

From 2000 to 2004 the volume of electricity generated from RES supported by the EEG increased from around 13.6 terawatt-hours (TWh) to 34.9TWh (BMU, 2004). During the same period the Act resulted in the volume of electricity generated from wind and biomass more than doubling, and brought about a nine-fold increase in electricity generated from PV systems in Germany. In 2004, wind generated power avoided the production of over 20 million tonnes of CO_2 (BWE, 2006). A total of around 70 million tonnes of carbon dioxide were already saved that same year, with 33 million tonnes of these being attributable directly to the EEG. The use of renewables prevented the emission of 83 million tonnes of CO_2 in 2005 (BMU, 2006a).

Using the latest data the Federal Environment Ministry demonstrates the increasing significance of RE for energy markets and the economy. These figures show that in 2006 the renewables' share of total electricity consumption in Germany rose to 11.8 per cent, compared to a 10.2 per cent share in the previous year. In 2006, 5.9 per cent of the heat and 4.7 per cent of the fuel for road traffic was from RES. In total, 7.4 per cent of the final energy consumption in Germany was supplied from renewable sources (BMU, 2006a).

In 2006, companies in the sector generated a turnover of €21.6 billion, up from €16.4 billion in 2005. A total of €6.5 billion was invested in new plants for the generation of power, heat and fuels. The renewables sector in Germany currently employs 214,000 people, and this figure is set to rise. This engages more people than in nuclear energy, hard coal and brown coal put together. By 2020 it is expected that 500,000 people will be employed in renewables.

Renewables make up a highly influential programme for growth, and by 2020 investments totalling €200 billion are expected (Friedrich Ebert Foundation, 2006).

The strength of the legislation is seen not just in the positive environmental and economic effects, but also at the level of the consumer. The EEG places almost no extra burden on home energy bills. The average power bill for a three-person household is approximately €52 per month (2004 figures). Only around €1.50 of this is for green electricity, produced through the EEG. The cost reduction with respect to environmental damage however, is at least €5.40 per household per month (BWE, 2006).

The effects of this policy will be felt increasingly through the advances in technology and improvements in administration that it engenders. The potential of RE will be seen increasingly in projects such as the 'Bioenergy village' of Jühnde in northern Germany. This award-winning project shows that 100 per cent of energy needs can be supplied by renewables. A bioenergy plant composed of a 700kW biogas installation and a 550kW wood chip heating plant provides electricity and heat for over 140 households. The heat is distributed via a short-distance heating pipeline with an overall length of 5500m. The plant is exclusively fuelled with local resources, which in turn is beneficial to the region (Eurosolar, 2005a).

The biogas is produced from the liquid manure of 800 cows, 1400 pigs, grass and other plants. It fuels a power station that generates 4,000,000kWh of electricity annually. In summer the generated heat is sufficient for heating and hot water, and dries the woodchips ready for the winter, when the wood chip heating plant becomes active.

Jühnde demonstrates that a 100 per cent energy supply from RES is perfectly feasible, depending on the individual RE endowment of the area concerned. The ecological, economical and regional advantages over conventional energy supply systems are clear, and this example further shows that rural areas can become self-sufficient in energy, saving money and making excellent use of by-products.

In looking forward, the German government has set itself ambitious goals: increasing the share of renewables to at least 4.2 per cent of the total energy consumption by 2010, at least 10 per cent by 2020, and to about 50 per cent by mid-century. It is also looking to expand biofuels use to 5.75 per cent of fuel consumption by 2010. While these types of targets are talked about with increasing frequency by various governments, Germany has demonstrated that it is not only committed to reaching these goals, but has the public support necessary to do so. This effort and belief is based principally on demonstrable successes, and the careful planning behind them.

Transition

The following quote from a report on jobs in the German renewables sector (Renewable Energy: Employment Effects, 2006) expresses the German perspective on domestic and international take-up of renewables:

> *A growing number of studies confirm that a significant contribution from renewable energy sources will be essential for solving or alleviating the energy supply problems pertaining to environment and climate protection, reducing import dependencies, avoiding conflicts over fossil resources, etc. Their significance is thus beyond dispute. However, concerns regarding the necessary financial expenditures during the expansion phase, and the resulting burdens for some economic sectors, are expressed. These concerns are becoming less and less sustainable because they usually originate from a short-sighted national perspective, the interests of individual economic sectors, or an assumption of permanently low energy prices. Recent developments in the global energy markets have demonstrated more clearly than ever that long-term international solutions are essential.*

With 2007 being the year that Germany holds both the EU and G8 presidencies, it will be a year in which many of the conflicting pressures on legislators are exposed. These pressures are going to intensify as the realities of climate change bite ever harder. The clash will grow between the need for a new energy system, and the desire of the old conventional energy sectors to cling to their investments and expertise rather than change with the times. Yet this is just one among many other major societal, economic and political transformations that are required, in the face of climate change. The fossil fuel energy system is the main culprit behind the rapidly increasing threat of climate change and the sooner the transformation takes place the better for us all. However, the transition stage is what we must grapple with in the immediate future. The transition from the old energy systems to decentralized renewable sources of energy is a phenomenon that can only take place effectively if all necessary agents are involved, from major global institutions and national governments, down to ordinary householders. Everywhere, there are examples of the beginnings of this change. Effective policies have allowed ordinary householders to take part in the transition in some places, and Germany is again at the forefront of this initiative.

5

Spain's Success

Spain, like Germany, has implemented a very effective feed-in law, with many positive gains, including a position among the world's top manufacturers of renewables technology. Again, its success in RE development is not due solely to this single policy, but a deeper commitment from the necessary agencies to respond intelligently to both its free energy resources, and its growing energy requirements. Spain's feed-in law is commonly considered to be the next most effective version in the world after Germany's, due in part to the explosive growth of Spanish RE industry and installed capacity, especially in wind and solar.

When the oil crisis hit Spain in the mid-1970s, the country had poor domestic energy supplies, covering only 28.6 per cent of its needs. The response was an attempt to diversify primary energy resources and develop energy efficiency, prioritizing the promotion of new energy types. Despite a growth in the supply side of renewables, the demand side also grew due to a failure to stabilize consumption, and hence the percentage of renewables in the mix stayed more or less the same between 1998 and 2003. Today, Spanish dependency on oil and gas is around 99 per cent and 97 per cent respectively, and primary energy dependence is almost 80 per cent. The need for increased energy autonomy is a clear driver for RE expansion in Spain.

An examination of the Spanish version of the feed-in system shows some marked differences from that implemented in Germany, and the scheme is still under ongoing development, which can be considered a major plus point in a feed-in system. In November 2006, Madrid hosted the Feed-in Cooperation's third workshop, at which they demonstrated the attention to detail that goes into the design of their system. The detailed section on Spain's policy development demonstrates, as in the German case, that increasing RE penetration is only possible through careful and flexible policy planning. Ragwitz and Huber (2004) produced an excellent study comparing the German and Spanish systems. This section is based on their findings, and demonstrates the level of divergence possible between two successful feed-in systems.

DEVELOPMENT OF SPANISH RE POLICY

The contribution of national policy to the development of the renewables market began in 1980 with the 82/1980 Energy Conservation Law. This began the process of introducing and refining the key legal guarantees of a feed-in policy, involving guaranteed access and price, and purchase contracts with the utilities. The Ministry of Energy and Industry set the price annually by Order in the first law, but did not determine contract length. Under a new electricity law in 1994, prices were instead set by Royal Decree. This law also developed the provisions for renewables, specifying a minimum of five years for purchase contracts.

In the 1997 Electricity Law, a target of 12 per cent primary energy from RES contribution by 2010 was established. The special regime conditions from this law were developed in a 1998 Royal Decree, which improved the investment environment. In 1999, a Policy Plan for the Promotion of Renewable Energy (PFER) dictated the methods of reaching the 12 per cent RES target. These goals were taken into account when establishing premiums. In 2005, a new renewable promotion plan was introduced.

ELECTRIC POWER ACT 54/1997

This Act introduced the current tariff system, and differentiated between the average rate of electricity production and what the law labelled the 'Special Regime' for facilities using non-consumable RE as primary energy, such as cogeneration, biomass or any other kinds of biofuels with installed power not exceeding 50MW. Further, it established the rights and responsibilities of RE producers, most notably the feeding-in of the energy produced into the electric grid, and the payment of a premium for this energy. These premiums were considered to be diversification and supply security costs of the power grid. Importantly, this law introduced competition in both generation and supply, with administrative licences allowing for market entry. The market was scheduled to be liberalized in 2007, but that was brought forward by Royal Decree to 2003. The liberalization of the energy markets was the most recent transformation with effects on RE installations. In comparison with the developments in other countries, this is relatively advanced in Spain.

The two basic issues to be addressed when developing RE sources in a liberalized electricity sector are how grid operators are to accept electricity supplied by RE producers, and at what price. The market price must be supplemented with a premium established in the corresponding regulations. As in the German EEG, this extra sum should reflect the social and ecological benefits of RE, provide an adequate investment return from RE generating installations in special regimes and increase investment security for RE installation projects.

Through this Act, the guarantee of quality and supply to customers replaced the traditional concept of public service. The economic and technical management of the system is performed by private enterprises, therefore making the operation of the national electricity system, both production and distribution, no longer a state-owned public service.

State planning was limited to transmission installations and is no longer effective for investments in electric companies. Unrestricted entry to electricity generation is acknowledged and organized under the principle of free competition. A wholesale market became the basis for sales and payment. Transmission and distribution were opened up through generalized third-party access to the grid. Ownership of the grids no longer guarantees exclusive use. Payment levels for transmission and distribution services will continue to be set by the government, which avoids the potential for monopoly abuse arising from the existence of a single grid. A transitional period was established for the liberalization of electricity supplies, whereby all consumers would gradually acquire the freedom of choice of supplier over a period of 10 years.

ROYAL DECREE 2818/1998

The Royal Decree 2818/1998, on the production of Electric Power by Facilities Supplied with Renewable Energy Sources, Waste and Co-generation, has been replaced by Royal Decree 436/2004. The 1998 Decree is worth looking at because the 2004 Royal Decree is based on and is very similar to RD 2818, so that understanding the former makes it easier to understand the latter; and also because its enforcement represented the greatest development in RE in Spain.

The Decree regulated the requirements and procedures of the special scheme, the registration procedures for the facilities in the corresponding registry, the conditions of energy delivery and the applicable economic scheme. For RE, it established, among other issues, the right of producers to incorporate all electric power produced into the national grid, and their entitlement to be paid a price on the wholesale market plus a bonus or premium. In addition, it established the initial values for these premiums and their annual updates, taking into account the variation of the average price of electricity sales. It also established a revision every four years in accordance with the evolution of the electricity power price on the market, the inclusion of RE to cover the demand and the technical management of the electricity grid.

Following the introduction of these acts, installed wind capacity exploded between 1998 and 2003, leaping from 834MW to 6235MW. The system was modified to improve management of the growing installed capacity of RE, and in response to criticism over increased control expenses (from intermittent weather-dependent technology) as well as to improve the predictability of the annual price adjustments to reduce investment risk.

ROYAL DECREE 436/2004

Royal Decree 436/2004 addressed these issues to some extent, and introduced several new features, including a guarantee on the percentage coupling of RES payments with the average electricity tariff (AET). There was, however, the introduction of a declining remuneration during the lifetime of an RE plant. The Decree, which establishes the methodology to update and systematize the legal and economic framework of the electric power production activity within the special scheme, consolidates the regulatory framework established by Law 54/1997 on the electricity sector for producers operating in the special system. Further, it provides additional incentives for the market-based tariff option, and mandated a production forecast for RE plants of over 10MW. Variable financial penalties were introduced for deviation from these forecasts. The fines were dictated by the choice of payment method and the RE technology in question.

The 2004 Decree helped to make the legal and economic framework for electricity generation under the special system more stable and predictable, and established a system based on the free choice of the producer, who can decide between either a regulated tariff or sale on the open market. RE producers can choose, for periods of not less than one year, the option that suits them best. Option 1: Sale to the distributor at the regulated tariff, which is the same for all scheduling periods, calculated as a percentage of the yearly average tariff as defined in RD 1432/2002, which approved the methodology for determining this tariff. Option 2: Free market sale, through the bidding system managed by the market operator (OMEL), the bilateral contracting system or forward contracting system (or both).

The price is set by the market, or negotiated by the parties in the case of a bilateral contract, plus an incentive and a premium for the power guarantee, like other producers under the ordinary system. The incentive for participation in the market and the premium are calculated as percentages of the yearly average tariff (defined in RD 1432/2002, which approves the methodology for determining this tariff).

Regardless of which payment system producers choose, the intention of this Decree is to grant the titleholders of the facilities under the special scheme a reasonable level of payment for their investments, and also to grant consumers an allotment of the cost ascribable to the electric grid. However, participation in the market is encouraged as this involves less administrative intervention when setting the electricity prices, in addition to a better and more efficient assignment of the grid costs, particularly as regards the management of the alternative routings and supplementary services.

Revisions of the 2004 Decree are carried out every four years, after 2006, in order to monitor and revise the mechanisms and prices as necessary, as well as the development and cost-effectiveness of RE technology. Other effects, such as the share of renewables in the national energy mix are assessed, and also are the effects of this on the technical and economic management of the electricity generation and supply system. The expansion of electricity from renewables in a national grid is a key question to address through any support scheme, but especially for feed-in laws as they normally give rise to faster expansion, and for more technologies.

Scope of application of the Royal Decree 436/2004

a) Self-producers using CHP or other forms of production associated with business activity other than electricity generation (provided the plant has a high energy efficiency).
b) Installations using renewable non-consumable sources of energy, biomass or biofuels. The categories are:
    b.1.   Solar.
        b.1.1. Solar photovoltaic.
        b.1.2. Solar thermal for electricity generation (with the possibility of using natural gas or propane: up to 12–15 per cent of electricity production).
    b.2.   Wind power.
        b.2.1. Onshore wind power.
        b.2.2. Offshore wind power.
    b.3.   Geothermal power and ocean power.
    b.4.   Hydroelectric with power $\leq$ 10MW.
    b.5.   Hydroelectric with power > 10MW and $\leq$ 50MW.
    b.6.   Biomass/energy crops or wastes from agriculture and forestry.
    b.7.   Biomass/biogas/sewage sludge/controlled landfill gases.
    b.8.   Biomass/industrial installations in the agriculture and forestry sector.
c) Installations recovering energy from waste material (MSW and others).

MAIN FEATURES OF THE SPECIAL SYSTEM

The electricity distributor has an obligation to buy electricity produced under the special system (provided this is technically possible) at the price set in RD 436/2004 and the regulatory body, the National Commission of Energy (Comisión Nacional de Energía or CNE), performs settlement of costs incurred under the special system by reimbursing distributors who have paid the prices, premiums and incentives laid down in RD 436/2004. The costs of electricity generation under the special system are taken into account for the annual calculation of the tariff, together with other costs: costs of generating electricity in the ordinary system, permanent costs, competition transition costs, transport and distribution, commercial management, diversification and security of supply (nuclear moratorium; second part of the nuclear fuel cycle). This allows the additional cost of the special system to be met by electricity consumers in a way that is proportional to their electricity consumption, an important factor in pricing mechanisms.

Forecasts for feeding electricity to the grid:

Decree 436/2004 obliges operators of RE installations (> 10MW) to provide the distributor with a forecast of the electricity they intend to feed into the grid at least 30 hours before the start of each day. Penalties are established for deviations.

Cost of deviation:

The cost of deviation will be 10 per cent of the average electricity tariff applied to the difference between the forecast and the electricity measured (when the permitted tolerance is exceeded – the tolerances are 20 per cent for solar and wind power, and 5 per cent for the rest). The cost of deviations for installations opting to sell directly to the market will be the same as that applied to installations operating in the ordinary system. The obligation to make forecasts and the penalties for deviations improve the functioning of the system and the quality of the electricity fed into the grid. The tariffs, premiums and incentives set out in RD 436/2004 were designed to be reviewed in 2006 and every four years thereafter. This would apply to new installations, not existing ones. A transitional period was established for electricity producers operating under the special system defined by RD 2818/98, lasting until 1 January 2007.

SUMMARY OF THE MAIN CHANGES IN THE ROYAL DECREE 436/2004

- Consolidates the system of support for RE already in effect, which is based on the guarantee to buy all the electricity produced at a price above that of the market (fixed rate or market average plus a premium).
- Makes the prices for electricity under the special system more predictable, as the prices, premiums and incentives are determined as a fixed percentage of the average electricity tariff published at the end of each year, and which apply to the following year.
- Improves the payment for electricity generated in PV installations (the threshold for receiving the higher premium goes from 5kW to 100kW).
- The regulated tariff for electricity produced in PV plants of ≤ 100kW is 575 per cent of the average electricity tariff. Producers cannot opt to sell on the free market.
- Differentiates between onshore and offshore wind power, but the regulated tariff, premiums and incentives are the same for both. The tariff for instal-

lations of > 5MW drops to 85 per cent five years after commissioning of the plant.

- The premium for renewable electricity is set at 40 per cent of the average electricity tariff, with the exception of solar power (250 per cent), hydropower > 25MW and ≤ 50MW (30 per cent) and electricity from biomass plants using wastes from agriculture and forestry (30 per cent).
- The incentive to participate in the market is 10 per cent of the average electricity tariff for renewable electricity, waste-powered and CHP plants treating and reducing wastes.
- The regulated tariff for electricity from renewable sources generally drops after 5, 15, 20 or 25 years from when the installations were commissioned. The premiums and incentives, on the other hand, remain at a fixed percentage throughout the useful life of the plant (with the exception of the premium set for solar power).

BARRIERS FOR RES IN SPAIN

Financing is a key issue as the various RE technologies themselves differ greatly in terms of profitability, due mainly to the variable support schemes in place, as well as market and technological development. Technologies with low market presence confront more barriers, including disinterest and passivity of local authorities, insufficiently trained personnel, few specialized companies, contradictory regulations in other sectors, and a lack of cross-sectoral policy integration. RE producers are required to pay for grid connection themselves, whereas in Germany the utilities are responsible for this, passing on the costs to customers through their billing system.

COMPARISON OF THE SPANISH AND GERMAN SYSTEMS

It is significant that despite the numerous differences between the Spanish and German systems, they share a very high level of effectiveness. As a result of the feed-in schemes implemented in these two countries, they have achieved the highest absolute increase of RE among all EU Member States. The systems are largely responsible for the increase of European generation of electricity from renewables, and also for the significant uptake of European wind capacity in particular in recent years. The feed-in systems have prompted major investments in renewables and are responsible for creating lead markets for RE technologies in both countries, as evidenced in, for example, the Renewable Energy Policy Network for the 21st Century (REN21) Global Status Report updates.

Both systems are characterized by both a relatively high static and dynamic efficiency. Whereas the high static efficiency is mainly based on the high invest-

Table 5.1 *Comparison of the main implantation characteristics of the Spanish and German feed-in tariff systems*

	Spain	Germany
Guaranteed duration of level of tariff	1 year[a]	Generally 20 years[b]
General duration of support	Long-term duration implemented in RD 436/2004, technology dependent (10 to 25 years)	Generally 20 years
Are the tariffs stepped?	No	Yes
Degression of tariffs	Set in a flexible way	Predefined (2–6.5% p.a.)
Implementation of burden sharing	Through system operator OMEL – leads to equal distribution among all electricity customers	Equal distribution among all electricity consumers
Premium tariff possible?	Yes	No
Direct access to the spot market in combination with FIT possible?	Yes	No
Supplemented by what kind of main additional support mechanisms	ICO-IDEA funding line, which provides with special conditions to investments in RE and RUE investments. In general, investment incentives, soft loans and tax incentives were defined under the 'Plan de Fomento de las Energias Renovables' (RES Promotion Plan), whose aim is to support RES investments with 13.1% public finance sources	Soft loans and investment incentives by the market incentive programme for biomass CHP, small hydropower, PV in schools. Tax incentives (reduction of income tax granted in the federal tax law especially for wind energy investments). Soft loans by a federal investment bank DtA (a relevant share of Germany's wind energy investments is financed by government loans)
Grid access	Guaranteed by the Act	Guaranteed by the Act
Costs of balancing power	Not to be covered by RES generator	Not to be covered by RES generator
Do specific tariffs for the following (sub)-technologies exist?		
Biogas	Yes	Yes
Offshore wind	Yes[c]	Yes
PV	Yes	Yes
Building integration of PV	No (only size dependent)[d]	Yes
Geothermal electricity	Yes	Yes
Solar thermal electricity	Yes	No
Ocean technologies	Yes	No
Refurbishment large hydro	No	Yes
Biomass CHP[e]	No	Yes
Renewable Biomass resources	Yes	Yes
Innovative technologies including fuel cells, microturbines etc	No	Yes

Note: a The possible annual change of tariffs in the new act R.D. 436/2004 is linked to the general electricity tariffs. Annual changes can therefore be only very moderate.
b Except hydropower (15 years for refurbishment of large hydro, 30 years for small hydro).
c In principle a special category but in fact the same tariff as for wind on-shore.
d No special tariff for building integrated PV exists but a size dependent differentiation (>< 100kW).
e Separate (additional) tariffs for biomass electricity production with CHP.

Source: Ragwitz and Huber, 2004

ment security offered by the two schemes, the high dynamic efficiency is reached through the early promotion of what are at present less mature technologies, such as solar thermal or PV. Both systems support a broad portfolio of RE technologies with specific tariffs and therefore provide the basis for a long-term and sustainable development of RES.

A further crucial similarity between the countries is that the FITs are supplemented by a broad array of additional support measures, in particular tax deductions on RE investments, soft loans with stable financing conditions and investment incentives (including subsidies and partial debt relief) for some selected technologies. This well-balanced policy mix increases the stability of the investments, and is one of the key success factors of the scheme. This point about providing a portfolio of complementary incentives cannot be made too strongly. Each and every paper which assesses renewables support schemes finds the same conclusion, that it is vital to provide an intelligent policy mix to help overcome barriers to RE deployment.

A comparison between the main parameters of the two schemes is summarized in Table 5.1 and relevant differences between the schemes are explained in more detail in the text below.

Guaranteed period of tariffs

Spain's relatively short period of guaranteed tariff levels creates, in theory, a substantially higher risk for investors. Therefore, higher requested internal rates of return on investments (and therefore higher interest rates) should have been observed in the Spanish market for RE investments compared to the German case in the past. In general, longer periods of guaranteed tariffs decrease investment risks and therefore the costs for society. However, the comparatively low level of tariffs at high growth rates for RE installations suggests that investment risks in Spain are not significantly higher, which could be caused by the generally very stable policy environment for renewables in Spain created especially by the Spanish Promotion Plan for Renewable Energy 2000–2010, or *Plan de Fomento de las Energias Renovables* (PER) (which has since been superseded by the 2005–2010 Renewable Energy Plan). Therefore, the high investment security observed in the two markets was more 'informal' in nature in the Spanish system than in the German one. Although this has not created major problems in the past in Spain, it could be a relevant issue for building up concentrating solar thermal (CST) capacities. Since CST is less mature as a technology than wind for example, risk is a greater element of any investment decision for this technology, especially when considered in combination with the significantly higher necessary investment costs for individual plants compared to other technologies.

Total duration of support

Both systems offer long-term support covering the lifetime of RE plants, although in Spain the exact level of support is in principle flexible. The general long-term stability of feed-in systems leads to a stable investment climate and technologically to the installation of high quality components. In both countries the feed-in laws are reviewed periodically.

Stepped tariff design

One notable difference between the systems is that the German system features a stepped design, and the Spanish one uses a flat structure in their tariffs. The stepped design of tariffs gives the opportunity to reimburse generation from RE in different bands of the (marginal) cost potential curve according to the actual generation costs. The main advantage of this approach is the lowering of the producer profits compared to a flat tariff design in the case of (very) efficient generation options. See Chapter 9 for further explanation.

Degression of tariffs

Another difference between the systems is the temporal degression of tariffs, which is implemented in the German case, applying the concept of technology learning. Rather than determining the future tariff structure beforehand, the Spanish system offers the flexibility of annual adjustments of the tariffs, which are determined year by year based on the current status of the market (tariffs might increase or decrease). Another difference between the two systems is based on the fact that degression affects only new investments in the German case but new and existing installations in the Spanish case. Therefore the Spanish system leads to 'overpaying' existing plants if the tariffs are increased and to financial underperformance for investors if the tariffs are lowered.

Existence of a premium tariff

One of the main benefits of a premium tariff, as used in the Spanish system, is that RES-E generation shows higher compatibility with principles of liberalized electricity markets. Furthermore the costs of social and environmental benefits of RES-E are directly measurable, which should have a positive effect on the social acceptance of the tariff system. Further differences with regard to the implementation of a premium FIT scheme instead of a 'fixed' tariff concern the interactions with the conventional electricity market, which are as follows:

- Both supply and demand on the spot market are higher. This is because under this scenario, no market separation between conventional and renewable power markets takes place. This can be essential if the power market is small, as the degree of competition rises due to the higher trading volume.
- As the revenues from RES-E are more uncertain than under a 'fixed' FIT scheme, investors will require a higher risk premium, leading to a lower RES-E deployment if not compensated by an additional premium (as in the Spanish system).
- In the case of an increasing conventional power price, producer surplus for RES-E generators rises too. On the contrary, by applying a fixed feed-in scheme the gap between the (rising) power market prices and RES-E generation costs decreases – instead of being constant in the case of a premium design – leading to lower costs for society. However, additional costs occur if the conventional power price drops.

Technology choices

The technologies supported by the feed-in systems in both countries with special rates exhibit some notable differences. Some of these choices are the obvious consequence of the available potentials for individual technologies, for example the non-existence of a separate tariff for CST systems in Germany. Other important differences with regard to technology differentiation are the existing support for ocean (wave and tide) applications in the Spanish system and the existence of separate tariffs or bonus systems for offshore wind installations, for building integrated PV systems, and for biomass CHP applications in the German system.

Cultural differences

The motivations behind Spanish and German uptake diverge to a notable extent (Peter Baz, Hugo Lucas Porta, by email). German citizens invest in renewables partly out of 'Ökologismus', meaning that they assume a level of personal sacrifice in support of environmental gains. On the other hand, Spaniards are said to be lagging behind in terms of awareness and commitment, and are motivated more by the investment returns made possible through the FIT system. The Spanish government's IDAE department (Instituto para la Diversificación y Ahorro de la Energía) deals with renewables and energy efficiency. It provides a great deal of material to help raise awareness, and a special TV comedy *Turn off the Light* is shown regularly to assist in this.

ACHIEVEMENTS OF SPANISH RENEWABLE ENERGY LEGISLATION

After over 20 years of support for RES through policy, with investment subsidies, soft loans and third-party financing, figures for 2005 showed that Spain had reached a contribution of nearly 6 per cent to primary energy consumption through RE. The 1997 Electricity Law set a goal of 12 per cent primary energy from RES by 2010 and the Renewable Energy Plan (PER) 2005–2010 established the goal of 30.3 per cent by the same year. Electricity production was made up of RE to the tune of 16.6 per cent by the end of 2005.

Between two of its largest wind companies, Spain has a 15 per cent share of the world market, with 550 companies working in the field. It is second only to Germany in the global wind market. Wind power has enjoyed the greatest increase on the back of policy incentives, with small hydropower in particular still held back by long-standing social and administrative barriers. Germany and Spain, importantly, have also been rated first and second in Western Europe for energy efficiency habits.

Projections of outcomes from the PER 2005–2010 include the creation of almost 95,000 jobs, and the saving of almost 77 million tonnes of CO_2 (compared to electricity production from natural gas).

FACING FORWARD

Spain's rapid growth in energy intensity (2,700,000 tonnes of oil equivalent added annually), its reliance on foreign energy sources (80 per cent over the last few years), as well as the given necessity of protecting the environment, have made the swift development of RE a chief concern. Along with energy efficiency, energy production and consumption affects national social, economic and environmental strategy, and has been addressed through the REP 2005–2010. It revises the Spanish Promotion Plan for Renewable Energy 2000–2010, maintaining the target of 12 per cent of total energy use to be supplied by RE by 2010, and incorporating other indicative targets. These are: 29.4 per cent of electricity to be generated from RES and 5.75 per cent of transport fuel to come from biofuels by 2010 (IDAE, 2005).

Thus far, wind, biofuels and biogas have made satisfactory progress, but small-scale hydro, biomass and solar are growing too slowly to meet the targets for 2010. In addition, EU directive 2001/77/EC, being transposed into Spanish legislation, governs the promotion of RE within national markets by 2010, adding further impetus to the Spanish RE plan. Further, directive 2003/30/EC sets the target for biofuels' introduction and use, which gives the Spanish agricultural sector an opportunity to diversify and take advantage of what is likely to become an enormously significant market.

The Spanish feed-in system is, at the time of writing, being debated in terms of refinements and improvements. It is likely to be altered in terms of introducing a cap on the amount paid via the premium tariff option. The Spanish Ministry for Industry, Tourism and Trade, and the Feed-in Cooperation can supply information on the status of the feed-in system.

6

Policies in the US

Promotion of renewables in the US is, due to the geography involved, a vast, complex area. At federal level, the existing Republican administration has been considered in environmental terms as little short of an international pariah. Support for renewables has also been limited, sporadic, uncertain and at times regressive. At state level, however, the story is different. The quantity and diversity of policies and programmes enacted is very significant, and demonstrates active engagement with climate change and energy security issues. DSIRE, The Database of State Incentives for Renewables and Efficiency (www.dsireusa.org) carries up-to-date information on these national policy measures, again using the complementary approach of renewables and energy efficiency.

Financial incentives for RE include tax incentives, grants, loans, rebates, industry recruitment, bond programmes and production incentives. Rules, regulations and policies for RE include public benefits funds, renewable portfolio standards (RPS), net metering, interconnection standards, extension analysis, generation disclosure, contractor licensing, equipment certification, solar/wind access laws, construction and design standards (including building energy codes and energy standards for public buildings), requirements to offer a green power product and green power purchasing/aggregation policies. For a detailed examination of individual policies and programmes for wind energy in 12 states see Bird et al (2005).

California is aiming for the highest level, in terms of having the most aggressive policies and incentives, and intends to become a world leader in wind and other renewables, as they were some decades ago. The reasons behind this state's success are explored in Gipe (1995), but in essence come down to a positive, proactive attitude towards renewables from the governor and other key actors, which helped in effectively transposing federal acts into state legislation. Therefore, no apology need be made for focusing on incentive schemes in that state, to see how they have become the biggest RE users in the world's sixth largest economy.

PURPA

Ironically, it was in fact America's invention of the Public Utilities Regulatory Policy Act 1978 (PURPA) (see p26), a federal proto-feed-in law, that stimulated record installations of different technologies in California. California's implementation of PURPA offered long-term (15–30 year) contracts at a fixed tariff for the first 10 years of facility operation. These were called 'Standard Offer 4' contracts. The policy did not really have an effect until the utility companies were forced to offer standard tariffs. When this was resolved and action began, 1500MW of wind power were brought online, providing 1 per cent of California's electricity for the last two decades – a good return for a state that is ranked only 17th in terms of US wind resources.

During its early years in California, PURPA's main effect was to increase industrial and commercial on-site cogeneration, but gradually project developers began to emerge and flourish in the state. The wind industry saw the greatest benefits. A 25 per cent California state tax credit for investments in wind power from 1980 to 1983 and an equivalent federal tax credit further bolstered development. Standard Offer 4 contracts helped to spur a sizeable market and manufacturing capacity developed for wind, geothermal, biomass, small hydro and solar technologies in the state. The generous RE endowment in California was as much a factor in the growth as the incentives to utilize them.

Other states also brought significant renewable capacity into operation during the 1980s, notably New York and Maine. In 2003, over 45 per cent of Maine's electricity supply came from RES (most from biomass and hydro), much of which was developed under the early PURPA contracts. Primarily as a result of state interpretation of PURPA as well as favourable tax incentives, 12,000MW of geothermal, small hydro, bio-power, solar thermal, and wind power generation facilities were constructed in the US during the 1980s. Of this capacity, 6100MW was installed in California alone – including 1600MW of wind, 2700MW of geothermal, 1200MW of biomass (Martinot et al, 2005).

PRODUCTION TAX CREDIT

The most effective federal-level incentive for renewables has arguably been the Federal Production Tax Credit (PTC). This is an inflation-adjusted credit which is paid per kWh of generation, based on the output of a qualifying facility during the first 10 years of its operation. In 2002, eligible wind energy generators earned an inflation-adjusted production tax credit of 1.8 cents/kWh.

Originally created under the 1992 Energy Policy Act, the Federal Production Tax Credit was initially available for projects installed between 1994 and 30 June 1999. The credit was subsequently extended to December 2001 and then again to December 2003. The impact of the tax credit on the wind energy industry is evident in the boom–bust cycle of development. Wind energy

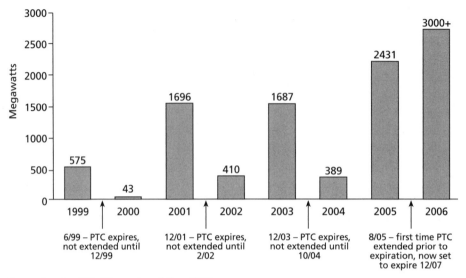

Source: American Wind Energy Association, 2005

Figure 6.1 *US wind capacity additions 1999–2006*

installations have peaked in years when the credit was scheduled to expire (i.e. 1999, 2001 and 2003) as developers naturally hastened to complete projects in time to take advantage of the credit. In contrast, development has slowed in the off years due to the uncertainty surrounding the extension of the PTC, and the lead-time necessary to plan and complete projects. The credit needs to be extended eight months ahead of expiration in order to give developers the lead time necessary to complete projects that can benefit from this scheme.

From 1999 until 2004, the PTC had expired on three separate occasions. Originally enacted as part of the Energy Policy Act of 1992, the PTC, then targeted to support just wind and certain bioenergy resources, was first allowed to expire on 30 June 1999. In December of 1999, again due to the efforts of various organizations, including the Union of Concerned Scientists (UCS), the credit was extended until 31 December 2001. The PTC expired at the end of 2001, and it was not until March 2002 that the credit was extended for another two years. Congress allowed the PTC to expire for the third time at the end of 2003. From late 2003 through most of 2004 attempts to extend and expand the PTC were held hostage to the fossil-fuel dominated comprehensive energy bill that ultimately failed to pass during the 108th Congress. In early October 2004, a one-year extension (retroactive back to 1 January 2004) of the PTC was included in a larger package of 'high priority' tax incentives for businesses (UCS, 2005b). The credit was set to expire on 31 December 2005, but due to the efforts of a coalition of clean energy supporters it was extended for two years as part of comprehensive energy legislation enacted in August of that year.

The legislation extending the PTC was included in a large package of tax incentives in the Energy Policy Act of 2005 (HR 6), which became Public Law No: 109–58. The PTC expansion was one of the few positives for RE in this energy bill. It provides a two-year extension (through to 31 December 2007) of the 1.5 cents/kWh credit for wind, solar, geothermal, and 'closed-loop' bioenergy facilities (adjusted for inflation, the 1.5 cents/kWh tax credit is currently valued at 1.8 cents/kWh). Other technologies, such as 'open-loop' biomass, incremental hydropower, small irrigation systems, landfill gas, and municipal solid waste (MSW), receive a lesser value tax credit (UCS, 2005b). The PTC has been extended ahead of expiration again during the last session of the 109th Congress in 2006, leaving it in place until the end of 2008.

Combined with a growing number of states that have adopted RPS schemes, the PTC has been a major driver of wind power development in recent years. The 2003 lapse in the PTC came on the heels of a near-record year in US wind energy capacity growth. In 2003, the wind power industry added 1687MW of capacity – a 36 per cent annual increase. With no PTC in place for most of 2004, US wind development decreased dramatically to less than 400MW – a five-year low. With the PTC reinstated, 2005 marked the best year ever for US wind energy development with 2431MW of capacity installed – a 43 per cent increase over the previous record, established in 2001.

Extending the PTC allows the wind industry to continue building on the momentum of previous years, but it is insufficient for sustaining the long-term growth of RE. The planning and permitting process for new wind facilities can take up to two years or longer to complete. As a result, many RE developers that depend on the PTC to improve a facility's cost-effectiveness may hesitate to start a new project due to the uncertainty that the credit will still be available to them when the project is completed (UCS, 2005b).

In 2005, Congress passed a two-year 30 per cent federal tax credit for residential PV systems. The effectiveness of this has yet to be determined, but this does at least provide an incentive for householders to engage with domestic RE production. This engagement has been instrumental in generating public support for renewables in Europe, something that provides a model for other regions in any RE strategy.

CALIFORNIA REBATE PROGRAM FOR PV

Between 1998 and the end of 2006, this programme put over 18,000 systems on Californian roofs. California's Emerging Renewables Program (ERP) has provided incentives for the purchase of four types of grid-connected RE generating systems: PV, solar thermal electric systems, fuel cells using renewable fuels, and small wind turbines. However, beginning 1 January 2007, the ERP will no longer provide rebates for PV installations. Incentives for installing PV systems on non-residential buildings and existing homes will be administered by the California Public Utilities Commission as part of the California Solar Initiative (CSI). Funding for integrating solar in new home construction will be administered by the California Energy Commission.

The ERP was offered to all grid-connected utility customers within the electric utility service areas of Pacific Gas & Electric Company (PG&E), Southern California Edison Company (SCE), San Diego Gas & Electric Company (SDG&E) and Southern California Water Company (doing business as Bear Valley Electric Service (BVE)).

From 1 July 2006, the rebate amounts were as follows:

- PV: $2.60/W for systems less than 30kW.
- Wind: $2.50/W for first 7.5kW and $1.50/W for increments > 7.5kW and < 30 kW.
- Solar thermal electric: $3.00/W for systems less than 30kW.
- Fuel cells using renewable fuels: $3.00/W for systems less than 30kW.

Rebates for eligible RE systems installed on affordable housing projects are available at 25 per cent above the standard rebate level up to 75 per cent of the system's installed cost.

Incentives received from sources other than this programme, such as other utility incentive programmes, a State of California sponsored incentive program, or a federal government sponsored incentive programme, other than tax credits, would reduce the amount of the ERP rebate by no less than 5 per cent to prevent total incentives from exceeding total system costs.

Participants in the ERP for PV systems could choose to receive the incentive as a capacity-based rebate in a lump sum as described above or as a performance-based incentive (PBI). The PBI was based on the amount of electricity generated by a system and was paid over a three-year period. System performance would be measured using a revenue-grade meter capable of measuring system generation in kWh. A total of $10 million was allocated to this pilot PBI programme for PV systems. The PBI level will remain constant for duration of the pilot programme.

- PV performance-based incentive: $0.50/kWh for three years.

There is no limitation on the size of an eligible system for the PBI option, but the funding cap for any system or group of systems at one site is capped at $400,000. In addition, the maximum funding available for all systems installed by any corporate or government parent is capped at $1 million.

The PBI programme could not be combined with other funding under the ERP, the Self Generation Incentive Program (SGIP), the Rebuild San Diego Program approved by the California Public Utilities Commission, or any other rebate programme funded with electric utility ratepayer funds. Twenty five per cent of the incentives received or expected from other sources will be subtracted from the amount of funds reserved for a system if these other incentives are from a utility incentive programme, a State of California sponsored incentive program, or a federal government sponsored incentive programme, other than tax credits (DSIRE, 2006a).

PUBLIC BENEFIT FUNDS IN CALIFORNIA

California's 1996 electric industry restructuring legislation (AB 1890) directed the state's three major investor-owned utilities (Southern California Edison, Pacific Gas & Electric Company, and San Diego Gas & Electric) to collect a 'public goods surcharge' on ratepayer electricity use from 1998 to 2001, to create public benefits funds for RE ($540 million), energy efficiency ($872 million), and research, development & demonstration (RD&D) ($62.5 million).

Subsequent legislation in 2000 (AB 995 and SB 1194) extended the programmes for ten years beginning in 2002, with annual funding of $135 million for RE programmes, $228 million for energy efficiency programmes, and $62.5 million for RD&D. In September 2005, the California Public Utilities Commission (CPUC) boosted energy efficiency funding to $2 billion for 2006–2008.

The California Energy Commission manages the renewables funds through four programmes:

* Existing Renewable Facilities Program – 20 per cent ($27 million/year)
* New Renewables Facilities Program – 51.5 per cent ($69.5 million/year)
* Emerging Renewables Program – 26.5 per cent ($35.8 million/year)
* Consumer Education Program – 2 per cent ($2.7 million/year)

The Existing Renewable Facilities Program is divided into two tiers: (1) biomass and solar thermal projects, which receive $20.25 million in annual funding; and (2) wind projects, which receive $6.75 million in annual funding. This programme supports the development and maintenance of existing RE projects (i.e. renewable projects that have already been constructed). This account uses a production credit mechanism based on the kilowatt-hours generated by a project.

The New Renewables Facilities Program supports prospective new RE projects that generate electricity. Once on line, the new facilities receive incentive payments for a maximum of five years, and like the Existing Program, incentives are awarded based on the number of kilowatt-hours generated.

The Emerging Renewables Program is being administered through a rebate programme. PVs, solar thermal electric, fuel cells that use renewable fuels, and wind turbines are eligible under this programme.

The Consumer Education Program provides funds to promote RE and help build the market for emerging renewable technologies. Consumer Education dollars are also used for tracking and verifying RE purchases under the Renewable Portfolio Standard.

Beginning in 2007, the California Energy Commission will manage $350 million targeted for incorporating solar energy in new residential building construction. It will use funds already allocated to the Energy Commission to foster renewable projects between 2007 and 2011. The CSI will provide a total of $2.9 billion over ten years for solar energy system rebates. Rebates for PVs provided under the Emerging Renewables Program will be handled under the CSI beginning in 2007. The state's goal is to install 1,000,000 rooftop solar panels on homes, businesses, farms, schools and public buildings over the next ten years to produce 3000MW, or the equivalent of six large power plants (DSIRE, 2006b).

RENEWABLES PORTFOLIO STANDARDS

RPS is the US name for the quota system. As described in the section on alternative support schemes, quota systems set percentage targets for the amount of renewable energy to be included in the power generation mix of a certain locale. To date, around 25 states have some form of RPS in place, with targets averaging around 20 per cent by 2020. Within this, there is some notable variation between states however, with California and New York setting the most aggressive targets, attempting to bring above-average amounts on stream ahead of the rest of the country.

When California's RPS was enacted on 12 September 2002 (SB 1078), it required retail sellers of electricity to purchase 20 per cent of their electricity from renewable resources by 2017, and was already the most aggressive RPS in the country. Because of perceived significant IOU (Investor Owned Utility) progress towards this goal, the Energy Commission and CPUC accelerated this goal of 20 per cent renewables to 2010 and set the state's 2020 goal at 33 per cent.

Eligible renewable resources include biomass, solar thermal, PV, wind, geothermal, fuel cells using renewable fuels, small hydropower of 30MW or less, digester gas, landfill gas, ocean wave, ocean thermal and tidal current. Municipal solid waste is generally eligible only if it is converted to a clean-burning fuel using a non-combustion thermal process. There are restrictions for some of these technologies.

Under the RPS, retail sellers of electricity are required to increase their procurement of eligible RE resources by at least 2 per cent per year, so that 20 per cent of their retail sales are procured from eligible RE resources by 2010. They are currently developing rules that will apply to investor owned utilities, and will later develop rules for electric service providers and community choice aggregators. Municipal utilities are ordered by the legislation to implement RPS programmes under their own direction.

The Energy Commission, in collaboration with the CPUC, has initiated a proceeding to implement the state's RPS. Pursuant to SB 1078 (2002), the Energy Commission must:

- certify eligible renewable resources that meet criteria contained in the bill;
- design and implement a tracking and verification system to ensure that renewable energy output is counted only once for the purpose of the RPS and for verifying retail product claims in California or other states; and
- allocate and award supplemental energy payments as specified in SB 1038 to eligible renewable energy resources to cover above-market costs of renewable energy.

The CPUC is addressing its responsibilities in implementing the RPS through a separate proceeding (Docket R. 01-10-24). The CPUC, in collaboration with the Energy Commission, is charged with:

- Determining market price referents for electricity from non-renewable sources. The IOUs will hold solicitations to purchase electricity from renewable generators, and bids above the referents may be eligible for supplemental energy payments from the Energy Commission.
- Establishing the process for the IOUs to follow in selecting the 'least cost' bidders of renewable energy that 'best fit' the IOUs' resource needs. IOUs will use the process to select winning bidders from their solicitations to procure renewable electricity.
- Implementing flexible rules for compliance with annual procurement targets. If an IOU fails to procure sufficient renewable energy, despite the flexibility, the CPUC will impose penalties.
- Establishing the standard terms and conditions to be used by all IOUs in contracting for eligible renewable energy resources. Parties had an opportunity to negotiate terms and conditions over the third quarter of 2003.

The California Legislature has charged the Energy Commission with developing a tracking system for implementing the RPS. In response, the Western Renewable Energy Generation Information System (WREGIS), a renewable-energy tracking system, is being developed jointly by the Energy Commission and the Western Governors' Association (WGA), with input from stakeholders (DSIRE, 2006c).

Despite California's efforts, which have not had the success hoped for, RPS policies in most states do not promote much beyond cheap wind installations. Solar PV gets little assistance except for in the New Jersey, Colorado and Pennsylvania versions, where they provide special 'carveouts' for this technology. The limitations of the portfolio in technology terms, as stated in the criticism of quota systems in Chapter 2, is one of the main basic errors in the system. The pro-quota systems argument concerning 'picking winners' is interesting therefore, as it appears that in fact quota systems do just that. The basic argument is that the government should not favour a technology through policy, but in general, the mechanics of the policy produce the outcome that the least-cost technologies will be supported first. This automatically pre-selects the lowest-cost option, which is onshore wind in most cases. Once all the wind sites are gone, the policy requirement should mean that the next cheapest technology is 'pulled through'. This is often offshore wind where that resource is available. With this in mind, winners are picked implicitly, if not explicitly, and other technologies are therefore held back by this system. This will certainly be addressed over time, as resources become scarce, and cheaper technologies proliferate and near their full installation capacity. The US has such a vast wealth of renewable resources that maintaining such a limited support scheme in the face of this is not only counter-intuitive, but betrays the dominant economic ideology at work. Only by reorienting this mindset, and prioritizing the utilization of the free energy that America has at its disposal, can it hope to break free of the fossil fuel dependence that has brought about such a host of international and domestic ills this last decade.

According to Wiser and Langniss (2001, p16), the most important problems experienced in US RPS design include:

- Inadequate attention to the relationship between the renewable energy purchase requirement and eligible renewable energy sources. For example, Maine established a 30 per cent RPS. Though this represents the highest RPS in the world, eligible resources include the vast majority of renewable energy and high-efficiency natural gas cogeneration in the New England region. Existing supply therefore far exceeds the standard itself. As a result, the RPS will do nothing to support new renewable energy development, and is unlikely to do much to support existing supply either.
- Selective application of the purchase requirement. Several US states only apply the RPS to a small segment of the state's market, muting the potential impacts of the policy. For example, in Connecticut the utilities that deliver energy to customers that do not switch to a new electricity supplier are exempt from the purchase requirement. Not only does this approach violate the principle of competitive parity, it also ensures that the RPS will have only a marginal impact, as the vast majority of customers have shown no interest in switching suppliers.
- Uncertain purchase obligation or end-date. Another common concern is the uncertainty in the size of the purchase standard and its end-date in some US states. In Maine, for example, the RPS is to be reviewed every five years. In Connecticut, when and how the RPS will end is simply unclear. Such uncertainty limits the ability of renewable generators to obtain reasonably priced long-term financing.
- Insufficient enforcement of the purchase requirement. Without adequate enforcement, retail electricity suppliers will surely fail to comply with the RPS. In this environment, renewable energy developers will have little incentive to build renewable energy plants. At best, the enforcement rules of a number of US RPS policies are vague in their application: these include those policies in Connecticut, Maine, and Massachusetts.
- Though of substantially less importance, still other states have failed to implement a renewable energy certificate system for easily tracking and

monitoring compliance with the RPS. States in this category include Maine, Connecticut, New Mexico and Pennsylvania.

This paper goes on to conclude, as does every paper this author has read, that growth in renewables is not the result of a single policy fix. The best results around the world are shown to be a product of a combination of policies and market mechanisms that support RE. In the case of Texas, the authors conclude that the factors involved were an effective RPS policy, a developing customer-driven market for green power, the wind power plans of electricity utilities, the federal PTC for wind, favourable transmission rules and an outstanding wind resource. However, without the PTC, it is likely that nothing of significance would have occurred. Further, they point out that the combined support mechanisms have only benefited wind energy. Two key points there: firstly to reinforce the conclusion that a mix of instruments is necessary just to have a chance against the highly subsidized conventional energy market; and secondly that with all of that in place, only one renewable technology has succeeded in getting a foothold in that enormous state.

Rickerson and Zytaruk (2006, p13) made some of the following points on RPS failures in the US:

- The focus on lower price has lead to a geographic concentration of wind farms, creating NIMBY (Not In My Back Yard) attitudes, where people oppose developments in their area. Further, the best sites for wind potential are taken first, and developers and investors may not be keen to consider building on lesser sites.
- RPS systems target near-market technologies, due to the lowest-cost technologies being 'pulled through' first, and leave technology markets to be supplied by foreign manufacturers. Deployment is often too slow and limited to encourage domestic production. As mentioned in the section on China, their government opted to avoid a feed-in tariff as it would speed up deployment too much, and they would not have time to set up enough domestic manufacturing to meet demand.
- Deployment rates under current regimes are not sufficient to mitigate climate change, which is one of the strongest arguments for having a support scheme in place, and expansion has fallen short of targets in some cases (UK, Nevada, US).
- Mono-price RPS market means windfalls for existing generators at good sites. Well-designed feed-in laws will guard against this.
- RPS systems favour oligopolies of large developers that can handle risks of investment in centralized generation; this then drives out small investors and discourages decentralization.
- No 'picking winners' means no portfolio diversity or encouragement of longer-term resources – thereby not making the system 'dynamically efficient'.

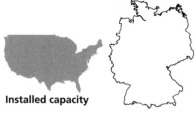

Area

Germany: 357,030km^2

Continental US: 8,154,157km^2
(~23 times larger, not counting Alaska)

Installed capacity

Continental US: 9149MW

Germany: 18,428MW
(~2 times larger)

Source: Perroy and Ryan (2005), in Rickerson and Zytaruk (2006)

Figure 6.2 *Landmass vs Wind Energy (MW) in Germany and*
Continental United States (2005)

• Administratively, RPS policies tend to be cumbersome and costly. This in itself is a major mark against them, especially when trying to find model RE-promotion legislation that can be replicated elsewhere. Many developing countries, for example, would prefer as simple a system as possible, due to the fact that they may have little or no experience with this kind of legislative instrument.

Figure 6.2 compares both landmass and installed wind capacity in Germany and the contiguous US. The figures were, as of the beginning of 2007: Germany, 11,603MW; and US, 20,621MW. The US is gaining, and the maps would alter slightly, but nonetheless Germany is still around 1.8–1.9 times larger in installed capacity.

The shortcomings of the RPS system are often highlighted, even in papers that are claimed to support them (Wiser and Langniss, 2001, for example). At the time of writing, there are clear currents of thoughts developing, suggesting that on the one hand, a move in the direction of FIT features is underway, and on the other, that a federal-level system will be implemented at some stage, whether feed-in or quota, presumably during the next administration. 'Banding' the RPS in the US, and the RO in the UK, has been mentioned on several occasions in personal correspondence and conversation. This would give support to all the different technologies, currently missing through the 'no picking winners' approach. However, the other key deficiencies of the quota system must be addressed, including rate of deployment, proper tariff differentiation, greater investor confidence and investment equality. If they fail to redesign the state RPS systems appropriately, they should be replaced with feed-in laws that, at the risk of labouring the point, do everything required of a support scheme for renewables.

7

Japan

Japan is, arguably as much by accident as design, among the world's leaders in solar technology. It is number two in the world for installed capacity of solar PV after Germany, but is a top importer of fossil fuels and, at government level at least, a nuclear energy enthusiast. It has a limited land mass and poor domestic energy supplies, many dense centres of population and an extremely energy-hungry neighbour across the sea in China. Japan has as much reason as any country to be concerned over its energy future.

Nuclear power – a highly questionable choice for a country prone to regular earthquakes – currently provides around a third of Japan's electricity, but the events at Hiroshima and Nagasaki have combined with an ongoing series of safety issues and scandals to create distrust of this technology among the public. However, this is the country that hosted the negotiations that led to the first emissions reductions treaty, the Kyoto Protocol, and consequently to a great increase in public awareness of the need for clean energy.

Despite the historical energy politics of the country, and dominating influence of the monopolistic utilities, there are reasons to believe that renewables will play an increasing role in the future of the nation's energy mix. The fact that Japan's own emission-reduction targets are likely to be missed may help to concentrate efforts.

With such a vast, established R&D and production base in solar technology in particular, and with global uranium reserves dwindling, optimists may suggest that it is only a matter of time before Japan accelerates its deployment of REs. However, energy politics within the country have hampered renewables development so far, and the targets for deployment are surprisingly timid – only 3 per cent of total energy supply by 2010.

Japan's first renewables R&D programme was launched in 1974, and in the following three decades the country passed the 1 gigawatt (GW) mark in installed capacity, with subsidy programmes and net metering schemes helping put solar panels on hundreds of thousands of residential, business and public buildings, halving the costs of production in the process. Japan's current share of worldwide PV production is over 40 per cent, and the sector, with many of the world's top companies represented, grew at 30 per cent annually for many years.

Voluntary net metering is one of the central factors contributing to Japan's PV development. In 1992, the ten major utility companies announced their programmes to buy surplus electricity from solar PV and wind at the same price as their electricity tariff. A special tariff pays ¥7/kWh at night and ¥30/kWh during the day. Significant market development was created by this measure, even though it was not a government instrument. It may even be perceived as a move by the utilities to head off any government regulation by setting their own terms for support of renewables.

The portfolio of measures that have been introduced in order to stimulate different elements of the solar supply chain have brought prices down to become competitive with conventional energy, thereby demonstrating once again that assistance is necessary at the early stages of market creation to compete with subsidized conventional energy.

Renewables in general have unfortunately encountered a variety of barriers in Japan, mostly from the utility companies, which have a monopoly, and from government itself. Utility companies have also tried to block wind integration, often on the grounds of intermittency issues. Geothermal has good prospects on these volcanic islands, but most sites are in protected national parks, and conflicts with hot spring businesses as well as more opposition from power companies have frustrated development.

In 1998 a revised programme of 'voluntary net metering' for commercial development of wind power was introduced. This paid a good rate (¥15/kWh) but was restricted to a one-year contract. Power companies have tried to avoid paying higher prices, so they offered to lower the purchasing price but offered a longer contract period. This proved to be a happy accident for wind power. With a longer contract period, the investment risk was lowered and development accelerated. When Hokkaido Electric Power Company (HEPCO) announced this programme in April 1998, they had only 3MW of wind power integrated. By the end of 1998, HEPCO had received inquiries for wind power development to the tune of over 500MW. This led to HEPCO limiting wind power introduction to 150MW in April 1999, again citing intermittency as the limiting factor. Other power companies have followed HEPCO in this way.

For the PV market, the focus in Japan has been on the grid-connected installations that have acted as a growth engine for the industry. The market for residential PV was assisted by the Ministry of Economy, Trade and Industry's (METI) Residential PV System Dissemination Programme, a successful scheme that has had a growing response over the past ten years, reaching an annual installation capacity of just over 200MW in the financial year 2003. A subsidy for domestic installations was first made available in 1994, and originally covered 50 per cent of the extra installation cost. In 1997, this switched to a fixed sum per kilowatt installed; 5654 installations were made that year. Since then, the subsidy has gradually been reduced, from its 1997 level of ¥340,000/kW ($3223/kW at current rates) to ¥90,000 ($865). Combined with net metering, this made PV systems even more attractive. In 2003 alone, nearly 53,000 domestic PV systems were installed in Japan. By March 2004, the residential programme had part-financed the installation of 169,000 systems, providing close to 623MW of residential PV. Local governments also provide financial subsidies (Jones, 2005).

In 2003, 262 local authority bodies gave financial support for PV. METI faded out the subsidy in 2005. It had played a fairly minor role in renewables promotion due to the relatively low rate, yet power industries were frustrated by the removal of the subsidy, and considered removing their own voluntary net metering schemes because of this.

Also in 2003, two major pieces of legislation were introduced. A revision of the Electricity Utilities Law, which governs power generation, electrical work and safety inspections, allowed new entrants to sell electricity to a broader range of customers, starting with midsize businesses that consume 500kW or more in (fiscal year) 2004 (April 2004 to March 2005), and extended to small businesses that consume 50kW or more in (fiscal year) 2005. This made it easier for independent producers with relatively little capacity to enter the electricity retailing business – a major shift in the sector. More than 20 energy supply companies have been set up as a result of the changes to the legislation. However, more than 99 per cent of the market is still monopolized by the existing power companies. Therefore, this legal revision has not yet helped to promote renewables significantly under the prevailing industry conditions.

Another significant legal development was the introduction of the Special Measures Law Concerning Promotion of the Use of New Energy by Electricity Utilities. It is a quota system, also known as a Renewables Portfolio Standard (see Chapter 6). The big utilities must therefore generate or purchase more RE from 'New Energy' which, under the law, covers wind, solar, geothermal, biomass and hydropower. Efforts to introduce a feed-in law were made, but the quota system was chosen.

Countering the effects of global warming is one of the central goals of Japan's present environmental policy, but METI only aims to bring on stream approximately 16.4 million tonnes of crude oil equivalent of new energy by 2010. This only sets a target for utilities to supply 1.35 per cent of total electricity from renewables by 2010. This has had the effect of becoming more a limitation than a target.

The price of PV in Japan has fallen dramatically over the years, with residential PV systems dropping in price by close to two-thirds in less than a decade. Similarly, the price of systems for public and industrial facilities has fallen, dropping below ¥1000 for the first time in 2001, and to ¥800 two years later. The promotion of these installations, under the NEDO field test programme, helped achieve the price reductions. NEDO, the New Energy and Industrial Technology Development Organization, which is funded by METI, started the field test programme in 1992 to demonstrate the effectiveness of introducing medium- to large-scale installations (Jones, 2005).

A 2005 report entitled 'PV market in Japan 2004/2005 – Current Topics & Future Prospects, and PV Activities in Japan', stresses the importance of PV within Japan's overall energy strategy. The government has worked to develop public awareness of climate and energy matters, and to highlight how solar PV can bring global, as well as personal, benefits. In support, there have been ongoing cross-media publicity campaigns from both national and local government on the benefits of PV in terms of wider environmental issues. The report concludes that increasing numbers of systems have gone to buyers who have a profound understanding of the wider implications of their purchase, even though the economic efficiency 'remains low'.

The report states that 'the PV system market, especially led by the segments of residential houses, public facilities, industrial and business facilities, is expected to expand and to grow to be a sustainable market in the near future by achieving cost reductions with the government's supports for research, development and introduction of PV systems.'

In its overview of the PV Roadmap towards 2030, the report notes that this period to 2030 will be a critical formation stage in the creation of a full-scale market for PV systems. A cumulative capacity of 100GW of PV in Japan is seen as achievable by 2030, by which time PV could meet 50 per cent of residential power needs, or 10 per cent of Japan's entire electricity supply. The PV price targets to be achieved by means of R&D are ¥23/kWh by 2010, ¥14/kWh by 2020, and ¥7/kWh by 2030 (Jones, 2005).

KYOTO PROTOCOL

As the country that hosted the 1997 Kyoto conference on climate change, Japan has always been one of its strongest advocates, ratifying the treaty in June 2002, and committing itself to reducing its carbon emissions substantially by the year 2010 – in Japan's case to 6 per cent less than 1990 levels. However, Japan's performance has been embarrassingly weak – carbon emissions have actually increased by nearly 8 per cent. At this rate it has little chance of meeting the obligations it signed up to, sending a rather negative signal to other Asian countries that are likely to become some of the biggest GHG producers over the next decade.

The nation's reputation for energy efficiency is in fact limited to the domestic sector. Industry is comparable in energy intensity to Germany. Households are reasonably efficient yet, according to the Institute for Sustainable Energy Policies (ISEP), account for only 5 per cent of the nation's emissions, and with increasing amounts of gadgets being introduced into the home, efficiency gains are thereby offset by increased consumption.

Some of Japan's geographic and cultural characteristics have also influenced the drive toward energy efficiency, producing cutting edge technological innovations in this area that also provide a lucrative export market. One of the key factors in this is the high retail cost of energy, causing heavy users such as the manufacturing sector to streamline energy use. Household appliances have become increasingly efficient, and the government has drawn up a law that would require manufacturers of air-conditioners, the heaviest drain on household electricity, to design units that consume 20 per cent less power by 2010. Japanese manufacturers, from steel, to cars, to electronics, are also some of the most energy-efficient in the world. This means making further cuts in emissions will cost Japan a lot more money per tonne of carbon than it will cost the US or EU. Japan already makes some of the world's most energy-efficient cars, and has persuaded increasing numbers of US drivers, the world's leading gas-guzzlers, to drive them. Its manufacturers are ahead of everyone else in developing fuel-cell, hybrid and flex-fuel engines.

Japan has shown great interest in Kyoto treaty components such as the Clean Development Mechanism (CDM) and Joint Implementation, which allow businesses in industrialized countries to buy carbon credits from countries that will exceed their Kyoto targets or are not bound by them, or to earn credits by investing in carbon-reducing projects in countries where it is easier and cheaper to achieve new efficiencies.

For example, the Tokyo-based power company Tepco is investing in a cassava-processing plant in Thailand, installing furnaces to burn the methane, a powerful GHG, produced by the plant for power generation. More than 40 such CDM projects have been approved by the Japanese government since 2002. It is now drawing up legislation to ensure this carbon trading is properly accounted, and a Dutch company is planning to establish the country's first 'carbon exchange', where carbon credits can be traded (Head, 2006).

Unfortunately, despite the measures implemented so far, it seems likely that Japan will fail to meet its Kyoto commitments. Carbon taxes, emissions caps and FITs have yet to be seriously addressed as remedies for this. At the time of writing, there are a growing number of alternative treaties either suggested or in development. Kyoto's failings, while beyond the scope of this work, are nonetheless highly pertinent to this subject, in that emissions reduction depends on less burning of fossil fuels for success.

Japan has had success with domestic PV deployment, albeit accidentally to some degree. However, to help the country increase RE production and export RE technology, and to foster an improved investment environment, a better instrument than the present quota system should be examined. The Tokyo Metropolitan Government may be the first to do so, with a feed-in model having been proposed. Liberalization of the energy sector may assist competition, and allow more entrants, but exploration of the renewables potential of the country across a range of technologies is more likely to truly accelerate under a national feed-in law, in conjunction with a strong commitment from government, private investors and the Japanese public.

8

The Developing World

Dan Bristow

Encouraging the spread of renewables in developing countries presents a differ-ent set of challenges to those in industrialized countries. The unique social, political and economic profile of each developing country will affect the feasi-bility of different policy measures. The most obvious of these in relation to FITs is the issue of whether or not there is a substantial national grid, without which the system cannot be directly applied in the same way as for developed countries. Using rural grid-connection as a dividing line helps demonstrate some of the profound differences in circumstance between developing countries – in much of sub-Saharan Africa only between 2 and 5 per cent of the rural popula-tion are connected to the grid, while in Thailand this figure is 98 per cent (Martinot et al, 2002).

More broadly, however, energy generation and access are closely linked to a range of economic, socio-cultural and geographical or environmental factors that can vary from country to country. Perhaps one of the most important of these is the relationship between energy and poverty. Improving access to energy is inextricably linked to tackling poverty. The availability of energy and the way in which it is generated has a profound impact on peoples' lives – their livelihoods, their access to water, their health and education, as well as gender-related issues. It impacts on the country as a whole – its economic productivity and its population levels. Yet the relationship is reciprocal, with poverty impact-ing on the generation and availability of energy – dictating the funding available for the development of large infrastructure at the macro level, and determining individuals' ability to pay for energy at the micro level. Clean, reliable and affordable forms of energy are essential for a country's development, but devel-opment and growth will affect that country's ability to access these forms of energy. Given the intimate relationship between energy and development, the governments of developing countries are faced with difficult decisions and significant challenges, but they are also presented with real opportunities in the form of renewables. The aim of this chapter is to consider briefly some of the challenges shared by most developing countries in promoting the uptake of renewables, and to look at some of the solutions that have been implemented across the developing world for both off-grid and grid-connected areas, where the focus will be on the implementation of FITs in Mauritius and China.

COMMON CONTEXTUAL FACTORS

In trying to promote renewables in developing countries one must contend with a considerable number of potential obstacles. The most pronounced of these is the lack of adequate funding. The large capital costs of renewables can make them prohibitively expensive. Although RE is by definition free, the costs of building, installing and maintaining the necessary infrastructure and technology can be considerable, particularly in relation to average incomes in poorer societies. In developing countries this is compounded by poor transport and communications infrastructure, which make the development of renewables projects more difficult and more costly. Similarly, the geographical dispersion of the populations of these countries adds to complexities and costs of renewables projects, with some small communities living in remote and hard to reach areas. There are also institutional and socio-economic factors – political instability, imperfect capital markets, low literacy rates, lack of trained personnel, and unsupportive or insufficient policy and regulatory frameworks (UNDESA, 2005). The latter is particularly pertinent.

In both developed and developing countries, the uptake of renewables relies on market creation and development through government policies and regulations. In this context the subsidization of conventional fuels needs to be challenged. It is estimated that of the hundreds of billions of dollars spent annually on subsidizing fossil fuels, 80–90 per cent is spent in developing countries (Sawin, 2004). It is clear that this will negatively impact on the affordability of renewables. However, creating a market involves much more than this. For grid-connected areas it means enabling private producers to cost-effectively supply the grid with the electricity they generate, and to be able to transmit this electricity using existing infrastructure at a reasonable cost.

Another crucial element for markets, whether grid-connected or otherwise, is capacity building (UNDESA, 2005). Developing countries often lack the domestic capacities to institute renewables projects (although this is gradually changing). Beyond the development of policy and regulatory frameworks, the design, development, implementation and operation of renewables programmes requires a level of knowledge and expertise. Similarly domestic investors, developers, entrepreneurs, private firms and consumers need to be able and willing to embrace renewable forms of energy. This is essential in the long term for the development of the country as a whole. Each country must, to some degree, be able to draw on domestic capacity (technological, institutional, economic, etc.) to maintain and increase the share of renewables in its energy mix. Intimately connected to this is the need for political, regulatory and market stability, without which the risks associated with investment in renewables can be too high. Despite these challenges there are a number of examples of renewables projects that are being successfully implemented in the developing world for both off-grid and grid-connected areas.

OFF-GRID

According to the International Energy Agency (IEA) 2006 World Energy Outlook there are still 1.6 billion people without access to electricity, and 2.5 billion people who use fuelwood, charcoal, agricultural waste and animal dung to meet the majority of their energy needs, in some cases relying on these sources for 90 per cent of their total household energy consumption. In many instances, because these fuels are burnt in poorly ventilated, enclosed spaces, they add to indoor as well as outdoor air pollution and encourage the spread of disease. The World Health Organization reports that indoor air pollution is responsible for the deaths of 1.6 million people every year (WHO, 2005). Aside from the health consequences, the inefficient and unsustainable use of these energy sources has other associated social, environmental and economic costs.

While constructing power lines and extending the national grid may appear to offer an obvious solution, it is often not feasible for technical and practical reasons. More than this, however, it is not always the best use of limited resources. Extending grid connections to remote rural areas can be prohibitively expensive, especially given the usually low population density and the nature of their energy usage. For example, rural areas in India with access to electricity still rely heavily on traditional biomass, particularly for cooking, where investing in electrical cooking appliances is too costly, and the power supply is typically unreliable (Bhattacharyya, 2006). As a result, many have been promoting alternative methods of satisfying the energy needs of the people in these rural areas, whether through providing RE for cooking and heating, or by providing electricity for productive uses (e.g. small-scale industry, agriculture, telecommunications, health, education and water services) through decentralized generation. There are a range of such energy solutions currently in use in various parts of the developed and developing world.

In developing countries:

- Photovoltaic (PV) and wind powered pumps can be used in water systems to pump clean water out of the ground or to pump water for irrigation.
- Anaerobic digesters can convert almost any organic material into a combustible gas. These digesters are being used in an estimated 16 million homes worldwide to provide energy for cooking and lighting (UNDESA, 2005). The by-products of digestion can also be used as fertilizers.

- Solar energy can be used for lighting, water heating and purification. Solar cookers have also been developed.
- Biomass stoves are used in cooking because they burn fuels at a high efficiency. Some stoves combine solar and biomass energy.
- Geothermal energy is being used for heating systems in a number of countries although its potential in remote, sparsely populated areas is limited.

For the purpose of decentralized electricity generation, solar home systems (SHS) can offer renewable domestic electricity production. PV modules are used to charge batteries which provide electricity throughout the day and into the night. In Mongolia, for example, a 100,000 solar ger (a ger is a wood-framed, canvas-covered circular dwelling used by herders) programme was started in 2002 to bring electricity to the nomadic herder families, and is now approximately half way to its target. Small-scale biomass gasification offers another source of clean energy production. By burning biomass at a high temperature in a low oxygen environment, gasifiers produce a gas that can be burnt in generators with a high efficiency and without producing smoke. This technology can help to supplement or replace the widely used diesel generators that currently supply mini-grids with electricity for tens or hundreds of houses in remote areas. Similarly, small-scale hydro, wind or solar power can be used to create hybrid mini-grids that run on a mixture of diesel and RE (wind diesel and PV diesel being the most popular).

Many developing countries have started to see renewables as part of a least-cost strategy for rural electrification. Historically, the promotion of these forms of RE in rural areas has either involved government subsidies or donor programmes. In China for example, the government invested $240 million over 20 months to subsidize the capital costs of their Township Electrification Programme, which electrified 1000 townships with solar PV, small hydro, and a small amount of wind power through a combination of SHS and mini-grids (National Renewable Energy Laboratory, 2004). Increasingly, however, new forms of financing have being used, including vendor-supplied credit, microcredit and rental models.

It has also been proposed that a variation on FITs could be used to bolster rural electrification (Aulich, 2006). The idea is to create hybrid mini-grids running on a mixture of diesel and small-scale renewables, with households operating microgenerators and being able to export electricity to the grid at preferential tariffs subsidized by the government. However, until this is operationalized the practicalities of its implementation remain unclear.

In Bangladesh the Grameen Shakti microcredit financing scheme provides affordable credit for SHS with the aim of helping the recipients to develop small-scale businesses (Grameen Communications, 2006). Families who participate are given a SHS and are encouraged to use the electricity to bolster their economic activity, either indirectly – by enabling people to work longer – or directly – by allowing people to develop new businesses, such as running mobile phones that they rent to members of the local community. Over time the participants use their increased income to pay for and ultimately own the SHS, which then becomes a practically free source of energy. This expansion of local economic activity can also be encouraged to spread further by increasing the local technological capacity. With the Grameen Shakti scheme for example, training is provided to members of the community to enable them to install, repair and maintain SHS. As noted above, this kind of capacity building is essential to the long-term sustainability of rural electrification programmes.

As the connection between poverty and energy access becomes more prominent, schemes like Grameen Shakti, which link rural energy programmes to both development and capacity building, must become the norm, and government efforts at providing energy to remote areas should concentrate on how alternative energy can help meet the development needs of the local population in this way. The focus must become one of catering to energy end-uses rather than simply providing electricity to these areas, so that the provision of renewable, reliable energy is linked to the improvement of health care and education, a reduction in drudgery and an increase in productive capacity. The key is to recognize and exploit the potential synergies between development and the provision of RE.

FITs IN GRID-CONNECTED AREAS

Both Mauritius and China have introduced FIT schemes and, although the contexts in which this has been done differ dramatically, both provide an insight into the flexibility of the feed-in system, as well as demonstrating the potential of the model for technological exploitation.

MAURITIUS

Mauritius has been cultivating sugar cane since it was first introduced to the island in the 17th century, and it continues to be one of the main pillars of the economy. Some 80 per cent of Mauritius's arable land is devoted to sugar cane, and despite a decline in its relative contribution to GDP, the country is still heavily dependent on the economic, social and environmental benefits of the sugar industry. One aspect of this is its role in the country's electricity system. The burning of bagasse – a fibrous by-product of sugar cane processing – has long satisfied the majority of the Mauritian sugar industry's energy needs, but with developments in technology and the long-term support of the government, including the provision of preferential or feed-in tariffs, the industry now exports 300GWh to the national grid (Ministry of Agro Industry and Fisheries, 2006).

Initially, the energy sector in Mauritius was run by small private companies providing electricity to consumers. This continued until 1952 when the government created the Central Electricity Board (CEB), which later became the sole supplier of electricity, paving the way for sugar factories to export electricity to the grid. The St Antoine Sugar Factory became the first sugar producer to do so in 1957 when it supplied 0.28GWh. From this inauspicious start, more and more sugar factories began supplying electricity until, by the late 1970s, this had grown to 25GWh. While the CEB was buying all the energy produced, the price at which it was doing so (0.6 cents per kWh) did not encourage investment, and sugar factories were simply selling the excess energy they produced. This meant that the supply of electricity from the sugar industry was unpredictable and intermittent, which made it difficult for the CEB to manage. This changed in the 1980s when a number of Power Purchase Agreements (PPA) were agreed, and a small number of sugar factories started to provide either continuous power (i.e. throughout the crop season) or firm power (all year round), with the latter being achieved by supplementing bagasse supplies with coal (Deepchand, 2002; Veragoo, 2003).

Since 1985, which saw the development of the Sugar Sector Package Deal Act (SSPD), the Mauritian government has implemented a number of policies to promote bagasse-based electricity generation. The SSPD aimed to improve the energy use of factory processes and encourage the storage of bagasse through the inter-crop season, but did not provide any incentives for the use of bagasse in energy production. This changed in 1988 with the introduction of the Sugar Industry Efficiency Act (SIE) which brought in a number of tax-based incentives for investment in generation, and for small planters to sell their bagasse to the larger power generating mills. Only a year after the enactment of the SIE the two private power producers renegotiated their PPA, and received a higher price per unit for the energy produced (Deepchand, 2002). Since that time, the price the CEB has paid for bagasse-based power has remained high with the result that the industry has made ongoing efforts to increase its power exports (Karekezi et al, 2005).

Seeing the positive impact the SIE had on the industry, the government decided to devise a strategy for optimizing the use of bagasse. In 1991, with expert assistance from the World Bank, the government produced the Bagasse Energy Development Programme (BEDP) to increase the share of bagasse-based power in the Mauritian energy mix. The main aims were to set up the necessary institutional framework and policies to encourage private investment in bagasse power production, and to use an investment programme to support this. To help achieve the latter, the World Bank agreed a $15 million loan with the Mauritian government, and the Global Environment Facility (GEF) fund provided a $3.3 million grant. In particular, the BEDP proposed the setting up of two firm power plants to be built next to existing sugar factories. These plants would use the bagasse collected from their neighbouring factories, as well as the surplus from smaller satellite factories in the immediate area to fuel their generators. When stores of bagasse ran out, coal would be used to ensure a continuous supply of electricity. By way of facilitating the efficiency of this system, attempts were also made to enhance the compaction and transport of bagasse, and to optimize the use of sugar cane for power generation (Deepchand, 2002).

Originally, the loan from the World Bank was intended to facilitate the building of one of the two aforementioned power plants, but this proposal encountered a number of difficulties. Specifically, the owners of the satellite plants questioned the price they were to receive for their bagasse, and suggested that an alternative, more favourable rate be set, and similar price negotiations were had between the private investor and the CEB regarding the pricing index for the PPA. Finally, a review of least-cost power development suggested that the proposed plant would be less cost-effective than the alternative (a combined cycle gas turbine plant). Although the methodology used to reach this conclusion was contested by both the private enterprise and the World Bank, the CEB disagreed with their analysis and refused to re-examine the conclusions of the review. Ultimately the plant was not built. However, the negotiations of the PPA led to the intervention of the government, which suggested that the price per kWh should be de-linked from CEB tariffs, and set in order to reflect costs (a form of FIT) (Deepchand, 2002).

As a result of the BEDP, the number of independent power providers has increased, and now practically all the bagasse produced in sugar processing is burned for electricity generation and cogeneration. There are currently seven sugar factories exporting bagasse-generated power during the crop season, and three independent power plants linked to sugar factories using a mixture of coal and bagasse to export electricity throughout the year; together these independent generators satisfy 45 per cent of the country's electricity needs (Ministry of Agro Industry and Fisheries, 2006).

This growth in bagasse-generated power has been closely connected to the setting of favourable tariffs, but interestingly this practice was never legally determined. It was instead achieved through the negotiation of the PPAs between the CEB and the relevant sugar factory, with the government playing the role of 'honest broker' (Karekezi et al, 2005). Contrary to what one might expect, this did not create market uncertainty. Arguably this is because the government has maintained its commitment to the development of bagasse generation for two central reasons – the importance of the sugar industry to the economy as a whole, and the high costs of fuel imports. This combination of pressures continues to drive governmental support for bagasse generation.

The upcoming changes to the European Common Agricultural Policy (CAP) will eliminate the preferential access to the EU markets that Mauritius has enjoyed until now. Faced with global competition on one hand, and a 36 per cent reduction in the price obtained for their sugar on the other, the sugar industry must diversify if it is to continue to thrive, and part of this diversification will come in the form of increased energy provision. The latest government strategy comes in the form of a Multi-Annual Adaptation Strategy (MAAS) which, with funding assistance from the EU, seeks to make the sugar industry more competitive. To this end, the MAAS aims to double bagasse-generated power by 2015 through the introduction of the latest technologies, and the optimization of bagasse use. Whether FITs are introduced conventionally, through a legal Act, remains to be seen.

CHINA

China's growing energy needs are well documented, and projections for the future show a continuing upward trend in energy consumption. With this growing demand already outstripping domestic supply, China is increasingly relying on imported oil to meet its energy needs, particularly for transport. In combination with worsening air quality and other forms of environmental damage, this has now encouraged the Chinese government to increase its commitment to RE. For the last few years RE development in China has seen an average annual growth rate of 25 per cent, and in 2005 only Germany invested as much as China in RE, with both countries investing $7 billion (excluding large hydropower) (Yang, 2006; REN21, 2006). This increased growth and investment was accompanied in February 2005 by the passing of the Renewable Energy Law (REL), which came into effect on 1 January 2006 and introduced FITs for the first time.

The REL was a clear expression of intent by the Chinese government. It enshrined in law a commitment to increase the share of RES in their energy mix, provided a number of financial incentives to encourage their development, and created a legal obligation for grid operators to purchase energy produced by licensed renewables projects. Despite this, the legislation did not do enough to ensure an adequate acceleration of the uptake of renewables. The main reason was that while it introduced FITs for biomass-generated power, it stopped short of extending this to other sources of RE. Although early drafts of the legislation included such a commitment in relation to wind power, this was dropped from the final version and replaced with a form of tendering system (Liu, 2006).

Under this policy, individual states put out tenders for wind power projects and offer the contract to the lowest-price bidder. A similar public tendering process has been in place since 2003 and the results have not been particularly promising. In practice it has meant that the majority of the tenders have been won by the country's five big power companies, which use the profits made in coal-fired power generation to subsidize their wind power projects – with some winning bids setting a tariff price lower than the cost of generation (Sun, 2006). This has led some to accuse the tendering process of causing harmful market volatility and uncertainty in the wind sector. Similarly the unsustainably low prices being set are said to be putting off foreign investors. However, the Chinese government's National Development and Reform Commission (NDRC) has argued that the pricing mechanism was designed to encourage the domestic wind power industry and, to this end, the government has issued a regulation stipulating that 70 per cent of wind power equipment should be procured from domestic companies regardless of the source of funding (Credit Suisse, 2006). Therefore, while FITs would offer the fastest market growth in wind power, the Chinese government appears keen to synchronize growth with the development of its own domestic renewables industry. An FIT would certainly lead to increased demand for foreign technology, much of which would come from Europe.

This is not the case for biomass projects, where the tariff has been set to match the local province price of desulphurized coal power (as of 2005) plus a fixed subsidy of 0.25 Yuan per kWh that will last for the first 15 years of operation. In order to encourage early development, the subsidy price is set to gradually reduce for projects agreed after 2010, dropping by 2 per cent each year. Often biomass generators are multi-fuel projects, which can accommodate renewable and conventional sources of energy. So to ensure that the FIT encourages only the development of RE projects, the legislation limits eligibility to projects with less than 20 per cent of energy being generated by conventional fuels.

As this legislation has been in effect for only a year, its impact is as yet unclear, but there are a number of potential stumbling blocks. Previous attempts by the government to encourage the growth in renewables have been undermined by a lack of effective and practical implementation and enforcement measures (Li, 2005). As noted in the discussion of wind power above, localized tendering processes have created market uncertainty. And with national legislation calling for solar, ocean and geothermal power tariffs to be set locally 'according to the principle of reasonable production costs plus reasonable profit', investors in these markets are faced with similar uncertainties, particularly surrounding tariff negotiations with local officials.

In addition to the lack of clear implementation guidelines, the current legal system appears ill-equipped to handle legislation like the REL. Not only are many laws hard to abide by because of the lack of supporting administrative acts and regulations but, more than this, the legal system is arguably ineffective. Chinese courts historically command little obedience, especially in those cases which affect government officials, with an estimated 40 per cent of court rulings failing to be enforced. Judges have also reported high levels of political interference, sometimes under the auspices of policy considerations, but often simply due to corruption (*Economist*, 2005). In such an environment, the potential impact of regulation like the REL is greatly reduced.

While the majority of investment in China has been in the development of small hydro and solar hot water projects (REN21, 2006) it is a country with great potential for large-scale grid-connected renewables projects. It has been estimated, for example, that China has a potential wind power generation capacity of 1000GW – more than two and a half times China's electricity generating capacity in 2004 (Junfeng et al, 2006; EIA, 2006). While the REL brought China closer to effectively promoting the uptake of renewables, more needs to be done if it is to realize its potential.

As is clear from the above, Mauritius and China face very different situations. In Mauritius the government's long-term commitment to the development and growth of the sugar industry was reflected in the role it played in facilitating negotiations between the CEB and the sugar industry, through which it fostered the introduction of FITs. The potential problems associated with developing capacity and finding adequate funds were ameliorated through international assistance. And now, the very fact that almost all the bagasse produced is burnt in generation and cogeneration shows how successful the government has been. In terms of the immediate future, changes in the trading relationship with the EU, and the associated restructuring of the sugar industry will take precedence, but in the longer term Mauritius must look to develop its solar and wind generation capacity. In doing so the government should seek to build on its previous success with FITs.

While China's rocketing economic growth has enabled it to invest heavily in rural renewables projects, the attempted introduction of an FIT system has served to highlight some of the institutional and socio-cultural challenges it still faces. The REL will undoubtedly have an impact on the development of renewables in China, but the degree to which it succeeds in its aims will depend on, among other things, the ability of the legal system to handle its enforcement. The point about China choosing to avoid FITs because they trigger such rapid growth, is effectively a back-handed compliment to the system. The downside is that few countries in the world need to make the transition from fossil fuels to renewables more than China, which is soon to become the world's greatest polluter.

CONCLUDING REMARKS

Generally speaking, increasing the share of renewables in the energy mix of developing countries involves addressing a number of interconnected factors that will vary according to the country under consideration. That said, the unifying factor is the need for governments to concentrate on creating the markets for renewables. In rural or off-grid areas this entails recognizing the synergies between development efforts, and programmes to electrify or to change energy use in these areas. Using renewables instead of traditional biomass for cooking and heating can offer profound quality of life improvements. Similarly, using renewables as part of electrification programmes can help encourage development while also limiting the harmful environmental impacts of conventional fuel use. Achieving this requires both commitment and concerted effort from governments. If subsidies continue to make diesel or coal cheaper than renewable alternatives, this will create on ongoing barrier to renewable deployment in rural areas. The governments of developing countries must draw on growing international experience to create the necessary policy and regulatory changes and attract investment – whether from international donors or from private investors (domestic or foreign).

85

For grid-connected areas, the lessons from China and Mauritius are mixed. In China the government has introduced the REL, which goes some way to accelerating the uptake of renewables, but weaknesses in the law and a lack of institutional capacity will undermine its effectiveness, and the government appears to have chosen to foster its domestic industry rather than ensure the fastest growth in the renewables market. In Mauritius, the growth in bagasse-generated power shows how the increasing renewables capacity can overlap with broader economic and social priorities. It is also an example of how the use of FITs can contribute to this growth.

For developing countries as a whole, FITs offer a proven method for speedy growth in the share of renewables in the country's energy mix. The extent to which FITs can be successful will depend on factors both within and beyond the government's control, but with continued commitment and international assistance it is a policy that could help developing countries leapfrog more developed countries, and place them at the forefront of the energy revolution.

Part 3

Implementation and the Future

9

Feed-in Tariff Design Options

Feed-in tariffs are by no means taken from a single mould and distributed to any country with immediate success. The table below shows the variety of potential components of an FIT, and how they have been selected in the design of different EU Member States' policies.

Klein et al (2006) examined the different design options for FITs in a comprehensive paper, which is recommended reading for any detailed investigation of

Table 9.1 *Feed-in tariff designs in the EU Member States*

Country	Purchase obligation	Stepped tariff	Tariff degression	Premium option	Equal burden sharing	Forecast obligation
Austria	X	X	-	-	X[a]	-
Cyprus	X	X	-	-	X	-
Czech Rep.	X (for fixed tariff)	X	-	X	X	-
Denmark	X	X	-	X (wind)	X[a]	-
Estonia	X	-	-	-	X	X (new draft)
France	X	X	X(wind)	X (new draft)	X	-
Germany	X	X	X	-	X[a]	-
Greece	X	X	-	-	X	-
Hungary	X	-	-	-	X	-
Ireland	X	X	-	-	X	-
Italy	X	X	X (PV)	-	X	-
Lithuania	X	-	-	-	X	-
Luxembourg	X	X	-	-	X	-
Netherlands[c]	X	X	-	-	X[b]	-
Portugal	X	X	-	-	X	-
Slovakia	X (for grid losses)	X	-	-	X	-
Slovenia	X (for fixed tariff)	X	-	X	X	X
Spain	X (for fixed tariff)	X	-	X	X	X

Note: a Austria, Denmark and Germany apply an equal burden sharing with advantages for electricity-intensive industries.
b In the Netherlands each electricity consumer contributes the same amount of money to RES-E support, regardless of the amount of electricity consumed.
c In the Netherlands no FITs are paid for electricity from RES-E plants that applied for support after 18 August 2006.

Source: Klein et al, 2006

potential legislation (see Table 9.1). This chapter summarizes the main points to consider from this examination, and shows some of the key considerations that will need to be addressed when researching best practices for implementation.

SUPPORT LEVEL AND DURATION

Tariff levels may be differentiated according to the variable costs of generating electricity from different RE technologies. Member States predominantly use this method to calculate tariff levels. By assessing costs, expected generation performance and estimated lifetime of the plant, an appropriate level can be determined. The factors affecting the final generation costs include:

- investment for plant construction;
- project-related costs such as licensing and planning;
- operation and maintenance (O&M);
- fuel (biomass and biogas);
- inflation;
- interest payments on capital invested;
- payment to investors.

Most of the EU countries with FITs apply the technology-specific option. Table 9.2 shows the remuneration levels and the period of guaranteed support in the EU countries. In the case of the premium option, the overall remuneration, which consists of the market electricity price and the premium tariff, is shown in order to be comparable to the fixed tariff. The tariffs are valid for renewables plants commissioned in the year 2006. Since some countries undertake a further differentiation of tariff level within one technology due to different framework conditions, ranges of remuneration levels are shown. In the case of stepped tariff designs (wind energy in Cyprus, Germany, France and in the Netherlands), the tariff level during the first year of operation is considered. The Czech Republic, Hungary and Portugal apply different tariff designs according to the time of day or season of the year. It is assumed that electricity generation in these countries has the same share in the different time bands. For the Czech Republic, Slovenia and Spain, all countries which offer a premium tariff, an electricity price of 5.56 € cents/kWh is assumed.

As with the design options, clear differences can be seen in the tariff levels. The time period for guaranteed support is highly variable, and ranges from one year to no limit, with an average of around 14 years.

Reimbursement rates vary greatly between countries and technologies, except where some countries have opted for uniform rates across all technologies, such as Hungary and Estonia. It would be interesting to chart the performance of the technologies in view of breaking a key rule of a good FIT; that of differentiating tariffs according to technology type. On the plus side for Hungary, they currently have no time limit for the rates.

Table 9.2 *2006 tariff levels and duration of support*

Tariff levels in 2006 (€ cents/kWh and duration of support for different technologies)[1]

Country	Small hydro	Wind onshore	Wind offshore	Solid biomass	Biogas	PV	Geothermal
Austria	3.8–6.3	7.8	–	10.2–16.0	3.0–16.5	47.0–60.0	7.0
	13 years	13 years		13 years	13 years	13 years	13 years
Cyprus	6.5	9.5	9.5	6.5	6.5	21.2–39.3	–
	no limit	15 years	15 years	no limit	no limit	15 years	
Czech Republic							
fixed	8.1	10.8	–	7.9–10.1	7.7–10.3	45.5	15.5
	15 years	15 years		15 years	15 years	15 years	15 years
premium	10.5	14.9	–	10.0–12.0	9.9–12.5	49.0	18.0
	15 years	15 years		15 years	15 years	15 years	15 years
Denmark	–	7.2	–	8.0	8.0	8.0	6.9
		20 years		20 years	20 years	20 years	20 years
Estonia	5.2	5.2	5.2	5.2	5.2	5.2	5.2
	7 years	12 years	12 years	7 years	12 years	12 years	12 years
France	5.5–7.6	8.2	13.0	4.9–6.1	4.5–14.0	30.0–55.0	12.0–15.0
	20 years	15 years	20 years	15 years	15 years	20 years	15 years
Germany	6.7–9.7	8.4	9.1	3.8–21.2[2]	6.5–21.2[2]	40.6–56.8	
	30 years	20 years	20 years	20 years	20 years	20 years	20 years
Greece	7.3–8.5	7.3–8.5	9.0	7.3–8.5	7.3–8.5	40.0–50.0	7.3–8.5
	12 years	12 years	12 years	12 years	12 years	12 years	12 years
Hungary	9.4	9.4	–	9.4	9.4	9.4	9.4
	no limit	no limit		no limit	no limit	no limit	no limit
Ireland	7.2	5.7–5.9	5.7–5.9	7.2	7.0–7.2	–	–
	15 years	15 years	15 years	15 years	15 years		
Italy	–	–	–	–	–	44.5–49.0	–
						20 years	
Lithuania	5.8	6.4	6.4	5.8	5.8	–	–
	10 years	10 years	10 years	10 years	10 years		
Luxembourg	7.9–10.3	7.9–10.3	–	10.4–12.8	10.4–12.8	28.0–56.0	–
	10 years	10 years		10 years	10 years	10 years	
Netherlands	9.7	7.7	9.7	7.0–9.7	2.1–9.7	9.7	–
	10 years	10 years	10 years	10 years	10 years	10 years	
Portugal	7.5	7.4	7.4	11.0	10.2	31.0–45.0	–
	15 years	15 years	15 years	15 years	15 years	15 years	
Slovakia	6.1	7.4	–	7.2–8.0	6.6	21.2	9.3
	1 year	1 year		1 year	1 year	1 year	1 year
Slovenia							
fixed	6.0–6.2	5.9–6.1	–	6.8–7.0	5.0–12.1	6.5–37.5	5.9
	1- years	10 years		10 years	10 years	10 years	10 years
premium	8.2–8.4	8.1–8.3	–	9.0–9.2	6.7–14.3	8.7–39.7	8.1
	10 years	10 years		10 years	10 years	10 years	10 years
Spain							
fixed	6.1–6.9	6.9	6.9	6.1–6.9	6.1–6.9	23.0–44.0	6.9
	no limit	no limit	no limit	no limit	no limit	no limit	no limit
premium	8.6–9.4	9.4	9.4	8.6–9.4	9.4	25.5	9.4

1 For the countries using a currency other than Euro, the exchange rate of 1 January 2006 is used [OANDA Corporation, 2006]
2 The maximum value for Germany is only available if all premiums are cumulated. This combines the enhanced use of innovative technologies, CHP generation and sustainable biomass use.

Source: Klein et al, 2006

Also, Slovenia presently has only one year as a fixed support period. This violates the general rule of providing medium-term investment security through supporting the installation for a good proportion of its working lifetime.

Germany has some of the most robust tariff levels, but is now closely matched, and in some places exceeded, by France. Greece has recently brought in some very commendable rates, with a guarantee period of 12 years across all technologies.

TARIFF REVISION

Tariff levels should be revised regularly in order to ensure that they are at an appropriate level to achieve the stated goals of the energy policy. Furthermore the power plant prices, which have a major impact on the electricity generation costs, may undergo unexpected changes due to varying 'input' prices for materials (e.g. for steel or silicon) or a technology breakthrough. The two main methods used to revise tariff levels are periodic revision and adjustment of tariffs, and capacity-dependent adjustment of tariffs. In addition, it must be determined whether or not the adjustment is just applied to new installations or also for the existing ones. A further question concerns whether or not the tariffs are adjusted to take inflation into account.

The revision process is key, as the outcome should be that all parties remain satisfied that they are paying enough, but not too much, for profitable operation and investment of RE installations. Designing a policy that is therefore flexible enough to accommodate changes in material inputs and technologies, while at the same time maintaining stability and low investment risk, is the challenge facing legislators. Thankfully, with so many countries now at work on their policies, there is plenty of scope for learning in this area.

The following sections will address some of the design options used in EU Member States.

PURCHASE OBLIGATION

Purchase obligations are obligations for electricity grid operators, energy supply companies or electricity consumers to purchase the power generated from RES. Most EU Member States have some kind of purchase obligation within their FIT, but there are exceptions. For example, there may be no purchase obligation for electricity offered on the spot market, or a purchase obligation only to the extent of electricity network losses.

For the former, Spain, the Czech Republic and Slovenia offer RE electricity sales directly on the spot market. In addition to the market price, the RE generators receive a premium per kWh of electricity. This concept of premium tariff design is used as an alternative to the fixed tariff design and the RES-E producers can choose one of the two options. While a purchase obligation is provided in these countries for the fixed tariff design, there is no purchase guarantee in the case of the premium tariff design. In Denmark, operators of onshore wind turbines (connected to the grid since 2003) must sell the generated electricity according to a premium tariff design, without a guaranteed purchase, and there is no alternative fixed tariff option offered.

Regarding the purchase obligation for network losses, in Estonia the grid operators only have to purchase the electricity from RES-E plants up to the level of their transmission and distribution losses. One reason for this rule is that not every network operator has a licence to sell electricity. Therefore those network operators without any licence cannot buy more electricity than the amount of their network losses. The losses in the grid are very low in times of low electricity consumption (for example during summer nights), and thus the purchase obligation is also low, which affects wind farms in particular. This legislation causes uncertainty for the investors. However, the Estonian government introduced a new draft amending the Electricity Market Act (RT I 2003, 25, 153), which provides a premium tariff design, allowing the RES-E generators to sell the electricity directly on the market. In Slovakia, similar legislation is applied where the law does not foresee a purchase obligation for the total amount of electricity from RES, but the operators of transmission and distribution networks have to purchase electricity from RES-E plants up to the level of their transmission and distribution losses.

One objection with respect to the purchase obligation is the fact that it does not represent market compatibility, because the electricity has to be bought independently from the demand. The premium option without a purchase guarantee is an attempt to enhance market compatibility. Typically such mechanisms to raise the market compatibility lead to an increase of tariff levels.

Stepped Tariff Designs

As shown previously, most EU countries apply differentiated tariffs for different RES-E technologies, in view of the varying generation costs. However, power generation costs may also differ between plants within the same RES-E technology due to the plant size, the type of fuel used, or the diverse external conditions at different sites, such as wind yield or solar radiation levels. In particular, the costs of electricity from wind energy vary significantly depending on the wind yield at a given site. In theory, the higher the yield (wind speed and duration), the lower the generation costs. This is because a large share of the generation costs is independent from the amount of electricity generated. The investment and installation costs as well as large parts of the operation costs and of the expenses for service and maintenance do not depend on the amount of electricity generated. If only one tariff level is applied for all wind turbines, the question is at what level to set the tariff. A high level makes many locations applicable for wind turbines, attracting many investors, and leading to a high exploitation of wind energy. However, plants at sites with a high wind yield can then be over-subsidized, again raising the prospect of public disaffection with RE payments.

The downside with a lower tariff level is that fewer sites can be profitably exploited, and so RE penetration is reduced. A similar problem is encountered with other technologies, with respect to differing generation costs and fuel inputs. With improved turbine design, lower wind-speed sites will be exploitable in future, and so the tariff level should be set intelligently. The most responsive way of addressing these issues is to use a stepped tariff design.

Stepped tariffs are differing rates paid within each technology type – as opposed to a flat tariff. Stepped tariffs can be designed according to location, plant size or fuel type. In the first category, location, the Netherlands, Portugal, Denmark, Cyprus and Germany all have differentiated tariffs for this, but all implement them in slightly different ways. Germany's approach, with the focus on the reference yield model, is described in Chapter 4.

For many RES-E technologies the specific electricity generation costs per kWh differ according to the plant size. The second group of stepped tariff designs takes this into account. Almost all EU countries applying FITs use different levels of remuneration according to the size of an RES-E plant. In most of these cases capacity scopes (for example PV plants with a capacity from 5 to 100kW) are used to determine the level of FITs. Portugal and Luxembourg use systems that are somewhat different.

Generation costs may also vary due to the type of fuel used, particularly with regard to biomass and biogas plants. Waste with a large biogenic fraction has a limited energetic potential. For countries with high RES targets, it will therefore be necessary to grow biomass specifically for the purpose of electricity generation, in order to use the full potential of biomass. However, such projects produce biomass (such as crops) with a higher price than the biogenic percentage of waste. In addition, producing biogas from animal residues is more expensive than the generation of landfill or sewage gas. These factors are taken into account by, for example, Austria, Germany, Spain and Portugal. In Germany the level of remuneration for electricity from biomass and biogas depends on different characteristics of the power plant as well as on the fuel type. Similarly in the situation in Austria, four different capacity ranges are distinguished. Furthermore the tariff level is increased if the biomass has not been treated before it is used as fuel and if the power plant fulfils certain criteria. In Spain too, the level of tariffs for biomass plants depends on the type of fuel used. Biomass from energetic cultivation, garden, forest, and agricultural waste receives a higher tariff rate than residues from industrial installations in the agricultural and forestry sector (for example the residues of olive production).

The table below summarizes the pros and cons of using a stepped tariff design.

Table 9.3 *Evaluation of a stepped tariff design*

Advantages	Disadvantages
• Differences in power generation costs due to the plant size or the fuel type can be taken into account. • Local conditions can be considered and reflected in the tariff level. • Not only the sites with most favourable conditions can be exploited. • Risk of over-compensating very efficient plants is minimized. • Producer profit is kept on a moderate level at favourable sites. Therefore the burden on electricity consumers is lower. • Higher electricity generation costs for example due to deeper water or a large distance to the coastline (in the case of offshore wind turbines) can be taken into account.	• The system can lead to high administrative complexity (e.g. for defining a reference turbine, as in Germany). • Many different tariff levels within the same technology may lead to less transparency and uncertainty for the investors. • If the tariffs for plants with a low capacity are significantly higher than for larger plants, it could be economically feasible to construct two small plants instead of a large one, even though larger plants might be more efficient. This decreases the overall efficiency of the system.

Source: Klein et al, 2006

TECHNICAL LEARNING FACTORS IN POLICY

A central outcome of RE policy should be to reduce costs through technological learning. This can be achieved through the provision of incentives for technology improvements and more efficient designs. The largest share of RE costs is made up by the price of the power plant itself and the installation costs. This is particularly applicable for technologies that do not require any expenses for fuel, such as wind power, PV, geothermal energy or hydropower. The price for power plants and the installation costs tend to decrease as a technology is applied due to the so-called experience curve effect or through technological learning. The resultant decrease in costs should be reflected by the support policy. This can be done by reducing the FIT level for new installations during the revision and adjustment of tariffs. Another possibility is a predefined degression of the tariff level, by a certain percentage per year, for new installations.

An experience curve describes the relation between the total costs that are associated with a technology (including labour, capital, administrative costs, research and marketing costs, etc.) and the cumulative output. In many industries it has been observed that with every doubling of the cumulative output the total costs per unit decrease by a fixed and predictable percentage, the so-called learning rate. The unit costs after cumulated output has doubled can be referred to as the progress ratio. A learning rate of 20 per cent (implying a progress ratio of 80 per cent) for example, means that the unit costs decrease by 20 per cent (down to 80 per cent) when cumulative output is doubled. The main factors that are made responsible for the reduction in costs are:

- learning process;
- economies of scale;
- technical progress;
- rationalization.

TARIFF DEGRESSION

As covered in the section on the EEG, German policy contains a degressive element, lowering the payments each year to new installations. The rate of degression, usually expressed as a percentage, should be calculated empirically, so that progress can be reflected accurately in the tariff structure.

The figure below shows the effectiveness of the German policy, through the decline of prices in this experience curve.

The price for wind turbines as a ratio of the annual electricity yield decreased from 80 to 38 € cents/(kWh/a) between 1990 and 2004. This implies a cost reduction of 53 per cent in total and an average learning rate of 5.2 per cent per year. This figure also shows that after stagnation of generation costs between 1990 and 1992, a strong decrease followed between 1992 and 1996. In the second half of the 1990s until 2004, costs decreased very little. The slower decrease is due to several factors. The development of wind turbines with a capacity above 1MW led to high costs for the turbine producers from 1996 on. Furthermore, steel prices have been increasing in recent years, and global demand for wind turbines has grown. However, technical improvements and more efficient solutions still led to a decrease in the specific electricity generation costs.

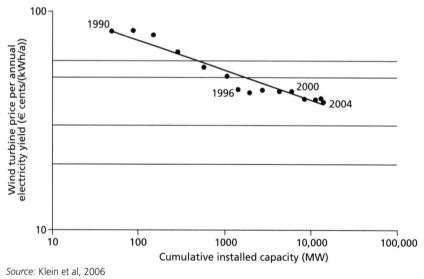

Source: Klein et al, 2006

Figure 9.1 *Experience curve of onshore wind turbines in Germany*

Table 9.4 *Evaluation of an FIT design with a tariff degression*

Advantages	Disadvantages
• Investment security. • Transparency. • Incentives to build a plant early in time, because the level of remuneration is decreasing along with the plant prices. • Incentive for technological improvement. • Lower burden on electricity consumers.	• If the degression rate is set for many years, the system is not very flexible in the case of varying technology prices due to structural changes (e.g. increased prices of steel or silicon). • It is difficult to set an appropriate degression rate, due to the difficulties in predicting technological learning, which is, for example, related to the cumulative amount of installed capacity.

Source: Klein et al, 2006

In the PV industry a similar development occurred, with even greater cost reductions. The price for a PV module decreased from $90 per Wpeak in the year 1968 down to $3.50 per Wpeak in 1998.[1] This implies a learning rate of 20 per cent. The prices of PV devices in Germany continued to decrease until the end of 2003. However, a fast growing PV market led to a shortage in the supply of the silicon used in PV module construction. This shortage was increased by a growing semiconductor industry, which also uses silicon. The silicon price rose from $30 per kg in 2003 to $60 in 2005. Approximately 10 per cent of the price of PV modules is made up by the price of silicon, and so the PV plant prices also increased in 2004 and 2005.

PREMIUM VERSUS FIXED TARIFF DESIGN

Tariff systems for paying RE producers can be separated into two types: an overall remuneration (the fixed tariff) or a premium, an additional payment on top of the electricity market price (the premium tariff). In the case of a fixed tariff design, producers receive a certain level of remuneration per kWh of electricity generated. In this case, the remuneration is independent of the electricity market price. In contrast, the development of the electricity price has an influence on the remuneration level under the premium option. Hence, the premium tariff represents a modification of the commonly used fixed tariff towards a more market-based support instrument. Currently, most of the European countries with feed-in systems opted for the fixed tariff model. Premium tariffs are applied (at the time of writing) in Spain, the Czech Republic, Slovenia, The Netherlands and Denmark. According to a new draft, amending the Estonian Electricity Market Act, premium tariffs are also considered for renewables support in Estonia.

The premium option shows a higher compatibility with the liberalized electricity markets than fixed FITs. This involves a better and more efficient assignment of the grid costs, particularly as regards the management of the alternative routings and supplementary services. The risk for the RE producers is larger in the case of the premium option, because the total level of remuneration is not determined in advance and there is no purchase obligation as is typically the case with the fixed option. Therefore the remuneration of the premium option has to be higher than that of the fixed tariff option in order to compensate the higher risk for RE producers (if the same investments in new installations are to be achieved).

Nevertheless, the higher support level also implies higher costs for the electricity consumers, especially if the remuneration levels of the fixed and premium options differ significantly, as we have seen in the Spanish example. One possibility to avoid these large differences and the extra costs for electricity consumers could be a premium varying with the electricity market price, as applied in Denmark, or a top limit for the overall remuneration paid in the case of the premium option. A bottom limit could be introduced as well, in order to compensate falling electricity prices. However, if varying premiums or limits are applied, some advantages of the premium option are no longer valid, for example the incentive to feed electricity into the grid in a moment of high demand (and a high price). On the other hand, the potential for operators of wind and solar power plants to feed the electricity into the grid at defined times is limited.

Table 9.5 *Evaluation of a premium tariff design compared to a fixed tariff design*

Advantages	Disadvantages
• More market orientated and less market distortion • More demand orientated • Provides an incentive to feed electricity into the grid, in times of peak demand	• No purchase guarantee, therefore less investment security • Most likely higher costs for electricity consumers, especially if the market price rises. • Operators of wind and solar power plants can hardly influence the time of electricity generation and therefore are not able to take advantage of feeding electricity into the grid at peak demand

Source: Klein et al, 2006

ADDITIONAL PREMIUMS FOR RES-E GENERATORS

Different premiums and incentives are applied in Germany and France for the following features:

- building-integrated PV;
- high efficiency of plants;
- regular electricity production.

If the electricity generation costs increase due to individual power plant designs, and these designs fit with the policy goals, it makes sense to pay an extra premium. An extra premium for high plant efficiency, which is paid in France for biogas or geothermal power plants, provides an incentive for plant operators to use the most advanced and efficient technologies.

REPOWERING PREMIUM

In Denmark and Germany an extra premium is paid to wind turbine operators if small, older wind turbines are replaced with new, larger models. Repowering has many benefits: with the same amount of wind turbines a higher electricity yield can be achieved; old turbines are usually built at sites with favourable wind conditions and might block them for new turbines; a repowering strategy may be used to improve the management of electricity grids; new and modern turbines are better adjusted to grid management; modern wind turbines typically rotate slower and run more smoothly than old turbines.

In Denmark, a removing certificate is granted to operators of wind turbines that replace a turbine with up to 450kW. This certificate brings an extra premium of up to 1.6 € cents/kWh for the first 12,000 full-load hours of electricity generated by the new wind turbine. To receive this premium, the old turbine must be decommissioned between 15 December 2004 and 15 December 2009. The premium is tied to the market price as the total of premiums and market price must not exceed 6.4 € cents/kWh.

DEMAND ORIENTATION

Electricity demand varies through time according to the time of day and the season, being higher during daytime and in winter. Lower temperatures and longer nights cause a higher demand for electricity during the winter months than in summer time. While this is typical in Europe, many hotter countries rely on air-conditioning throughout the day in summer periods, and recent rises in global temperatures are extending the demands on air-conditioning around the world, and so this curve is set to change over time. Some countries with FITs take the time of the day or the season of the year into account when setting tariff levels. For example, the Portuguese legislation provides different tariffs for day and night. Plant operators can decide if they want to receive the same tariff level independently of the time of day or if they want to receive a higher remuneration for electricity fed into the system during the day than during the night. Hydropower plants, however, are obliged to receive differing tariffs according to the time of day.

Table 9.6 *Evaluation of demand orientated tariff levels*

Advantages	Disadvantages
• Good system to take electricity demand into account • More market orientated than just one tariff level • Possibility to make RES-E generators more sensitive to electricity demand • Incentive to feed electricity into the system when it is needed most	• Higher administrational complexity than one tariff level • RES-E generators might not always know when the electricity demand is high • Does not make much sense for wind and solar power, because the operators cannot influence the electricity generation

Source: Klein et al, 2006

METHODS OF INCREASING LOCAL ACCEPTANCE

Portugal and Greece have instituted a policy whereby RE producers pay the local municipalities a percentage of their remuneration. This has the advantage of increasing local support, and even generating active participation, in addition to driving efficiency (and hence higher revenues for the local authority). On the other hand, it raises both administrative complexity and final prices for energy.

BURDEN SHARING

In countries where the electricity from new RES-E plants contributes significantly to the total electricity consumption, the distribution of the costs emerging from the support of RE is a crucial aspect of the FIT design. In most EU countries these costs are distributed equally among all electricity consumers by including them in the power price. However, distinct consumer groups are affected in different ways by the increased power price due to RES-E generation. Electricity-intensive industries in particular may ultimately have their international competitiveness affected by paying higher energy prices. To address this, some European countries have implemented a burden-sharing element of their FIT, depending on the consumer type.

In order to determine which electricity-intensive industry sectors are affected most by increased power prices due to RES-E generation, the following indicators can be applied, among others:

* Total amount of electricity consumption of a company.
* Annual costs of electricity consumption in relation to other parameters, such as the revenues, the total costs or the gross value added of a company.
* Voltage level of the grid connection. (Usually, a connection to a high voltage grid implies that a consumer uses a high amount of electricity.)

Since the use, and therefore the costs, of RE varies among Member States, the burden on the electricity consumers varies as well. This affects electricity-intensive industries more than other consumers. Burden sharing among all electricity consumers in Europe could be considered for solving the problem of international competitiveness. However, this requires coordination of the support systems for RE as well, otherwise a country could pay very high tariffs to the RE producers, knowing that the costs are distributed among all electricity consumers in Europe.

FORECAST OBLIGATION

For some types of RES the amount of electricity generated depends on external conditions like solar radiation, wind speed or the level of water in a river. An integration of the electricity from these RES into the power grids is much easier if the amount of electricity that is generated can be forecasted. The amount of water in a river is reasonably predictable and changes to this are quite slow. Therefore, the amount of electricity from hydropower plants is predictable. The integration of electricity from PV plants does not have great influence on the electricity grid, because the share of PV electricity is still very small. This is different for wind power, as the wind conditions tend to change very fast and the share of electricity from wind energy is very significant in some areas.

In some countries the operators of RES-E plants are obliged to predict the amount of electricity they plan to feed into the grid. For example, Spain applies a system with a forecast obligation. In the case of the fixed price option, only plants with a capacity of more than 10MW are affected by the forecast obligation. RE producers must report to the grid operator the amount of electricity they plan to feed into the system for each hour of the day, at least 30 hours before a day starts. Up until one hour before an hourly interval starts, it is possible to correct the predicted amount. If the penetrated electricity differs from the provision by more than 20 per cent in the case of solar and wind energy and by more than 5 per cent in other cases the operators have to pay a fee of 10 per cent of the reference electricity price for each kWh of deviation. For those plant operators who choose the premium option, the market rules are applied. Therefore they have to forecast the amount of electricity generated for all RE plants (not only the ones with a capacity of more than 10MW). For deviations a penalty of 10 per cent of the daily market price has to be paid. This legislation makes integration of electricity into the grid easier. Furthermore, it provides an incentive to improve the forecast quality, because of the penalty. It has to be stated, however, that RES-E producers have the option of compensating missing electricity from one wind farm with excess electricity from another. In addition, electricity from other types of RES can be used to balance the deviation.

CONCLUSIONS

As in the following chapter on general recommendations for best design practices, this key paper finds the same things apply in an EU-focused assessment of best practices:

- RES-E support requires continuity and long-term investment policy.
- Technology-specific tariff levels should be applied.
- Energy policy should provide mechanisms to ensure the penetration, and improve the integration of RES-E into the grid.
- A premium tariff option can be applied to increase market orientation.
- Tariff degression provides incentives for cost reductions.
- Stepped tariffs may be applied to reflect different power generation costs within the same technology.
- Extra premiums may help to reach policy goals.

This common list is also reflected in analyses of quota systems, demonstrating that despite the differences among them, the key goals of policy cannot be met without careful design, a long-term commitment and individual treatment of technologies.

NOTE

1 Wpeak is a measure of how much energy such a solar panel can produce under optimal conditions.

10

General Lessons for Effective Implementation

The crucial environmental, social and economic benefits of renewables have been proven to be delivered most effectively through a comprehensive, well-designed and implemented FIT system. The following points constitute a distillation of the lessons learnt from around two decades of European feed-in legislation experience.

- Provide tariffs for all potential developers and investors, including utilities, to encourage broad participation.
- Eliminate barriers to grid connection. Guaranteed access to, and strategic development of the national grid, are essential elements in developing investor confidence.
- Ensure that tariffs are high enough to cover costs and ensure a profitable operation.
- Keep tariffs to a level at which they provide just enough return on investment for profitable operation, but add no unnecessary burden to the end consumers, as this will erode support for renewables.
- Encourage technological development through annual rate decreases for newly installed RE systems.
- Provide differentiated tariffs to provide only the level of support that is necessary at the time for each technology.
- Design the system to be flexible. It is essential in the design of policies that adjustments (fine-tuning, but not wholesale changes or elimination of policies) can be made on a regular, predetermined time schedule if circumstances change. Governments must be able to address existing barriers as they become apparent, and new barriers as they arise. Policies also must be designed to allow developers/generators flexibility for meeting government mandates.
- Policies must match RE and carbon emission reduction objectives, and should take account of a country or region's variation in resource potentials, location, technology type, and timing.

- Provide financial security: guarantee tariffs for a long enough time period to ensure sufficient rate of return. The long-term certainty that results from guaranteed prices over 15–20 years helps companies to invest with confidence in technology, staff expansion and training, and the establishment of other services and resources on a longer-term basis. Obtaining finance from banks and other investors is easier, as they are assured a guaranteed rate of return over a specified period of time. Long time periods also help to provide enough lead time to allow industries and markets to adjust.
- Develop appropriate administrative procedures and streamlined application processes. Keep administrative costs and demands low. In general, FITs are easier to administer and enforce than quota systems. As with quota systems, policy makers are required to establish targets and timetables, and to determine which technologies are qualified (type and scale).
- Keep the structure of the system as simple as possible. Policies must be easy to implement, understand and comply with. Procedures of permission and administration, where necessary, must be as clear and simple as possible.
- Ensure public acceptance by raising awareness, and designing the system so that the costs to the customer are minimal. The German system is exemplary in this regard, adding only around one euro per month to the cost of RE deployment in the country through the EEG. Independence from state budgets is likely to help with positive public perceptions.
- Make policies both credible and enforceable. If policies are not credible and effective, or are not enforceable (or enforced), compliance will be diminished.
- Transparency is important for suppliers and consumers of energy and is necessary to avoid abuse. It facilitates enforcement, maximizes confidence in policies, and helps ensure that mechanisms are open and fair.
- If possible, a stepped tariff design should be implemented to reduce windfall profits, and therefore reduce costs for consumers (Ragwitz et al, 2005; Sawin, 2004).

Low investment risk is perhaps the most important factor, whether for a private householder wishing to install solar panels and sell some of the electricity back to the grid, for farmers using their farm wastes for biogas, or for corporate investors looking to build a large-scale wind farm. It is vital that any and all potential producers of RE should be catered for. Investment and planning security are assured by the specific design of the EEG, and should be key baseline considerations when transposing feed-in legislation into national or state law. Investment risk is greatly reduced through the combination of differentiated tariffs (type and scale) and legal guarantees regarding grid access, payment level and duration.

The technology-specific remuneration for RE electricity is a vital design criterion for successful RE deployment. It should be easily adjustable within a well-designed FIT system. Differentiated tariffs take account of the specific development level and cost-effectiveness of each technology. This is essential to attain a broad RE technology mix. The EEG's broad promotion approach sets remuneration rates according to the technology used, the size of the plant and, in the case of wind energy, on the age and size of the installation. Tariffs can be adjusted to reflect changes in technologies and market conditions over time. This flexibility was incorporated into the German system in 2000, and is now featured in other national feed-in laws as well. This approach can also help to influence the amount of new capacity coming online as desired.

In the case of the EU, problems have been encountered where some countries have fallen afoul of article 87 of the EU treaty concerning state aid. Both the StrEG and the EEG were challenged on legal grounds as constituting state aid, but the European Court of Justice found that the former did not constitute prohibited direct or indirect state aid, and the latter likewise. The key issue is that the government must not receive any funds from the scheme; the law should be set up as a legal arrangement between producers, grid operators, utility companies and customers.

Government commitment to develop RE markets and industries must be strong and long-term, with a clear goal of supporting accelerated development and proliferation of the various technologies, just as it has been with fossil fuels and nuclear power. It is unlikely that some form of international agreement to remove all subsidies for these conventional energy sources will be implemented, and so the support for renewables must make them artificially competitive with artificially cheap conventional energy. While this double subsidization may appear somewhat ridiculous, no politically realistic solution has yet been tabled.

The dominance of economic considerations shows one of the major systemic problems at work: the question of affordability of renewables is still being asked above all others. For those who work on survival strategies for the Earth's species in the face of growing human threats, for practical purposes cost becomes an irrelevant question, and certainly not the factor that should decide at what rate we take action to ensure the survival of the biotic component of the Earth on which we immediately depend. It would in fact be of major benefit to mandate the study of a course in Earth sciences for all key decision makers and their advisers, in order to understand clearly the mutual dependency of all species, and the basic conditions for life on Earth. We are not separate from nature, no matter how intensively we continue the transition to becoming a predominantly urban species. We cannot buy our way out of dependency on the natural world. The erosion of biodiversity, habitat loss and the collapse of ecosystems all have direct consequences for humanity, whether at local or global level – quite apart from the climate change factors which are accelerating this destruction. Science, technology, medicine and nutrition, to name four key areas, learn, borrow and take from nature every day, and have done throughout our existence. Nature has always provided answers to our problems, and supplied our needs and wants, but it can only continue do so while in a state of health.

The next two decades will almost certainly see a forced shift in policy drivers around the world, as the realities of climate change manifest themselves in increasingly dramatic and expensive ways. The message that we cannot afford *not* to prevent climate change has yet to penetrate significantly the collective political psyche, even as individual countries, states and provinces take some actions of their own. This will change in the near future, but it is still not a straightforward process. The brief geography of environmental decision making that I have noted has shown a clear trend towards politicians making decisions based on the dominant energy resources of their constituency. Areas with vast coal reserves, for example, tend to be represented by avid proponents of coal use who are less enthusiastic about renewables – another example of artificial barriers to renewables.

Just because such barriers will perforce crumble as we run short of fossil fuels and uranium, this by no means ensures that we could, under such circumstances, make the transition in time to preserve the self-regulating functions of the climate system. The combination of renewable energy and energy efficiency is the only safe, realistic choice we have for ensuring our future, in terms of environmental protection certainly, and also the avoidance of pollution, conflict over resources and nuclear proliferation. With a proven system already in existence for global acceleration of renewable energy deployment, from which the above lessons have been learnt, the policy wheel need not be reinvented, and we need not wait to begin the rapid transition which is vital for all of the above reasons.

11

Organizations Aiding Renewable Energy Deployment

In the continued absence of an international RE agency to take on this task at the global level, there exist numerous national and regional organizations that can provide assistance to governments on the design and implementation of feed-in tariffs, and help to share knowledge, expertise and best practice approaches.

The German Federal Ministry for the Environment, Nature Conservation and Nuclear Safety (BMU) will arrange meetings to provide consultation to interested parties on the implementation of FITs in other countries, states and provinces. They have great expertise and resources in this area, and would be a recommended first point of contact for FIT enquiries. See www.bmu.de.

EREC is another organization providing information and consultancy on RE for political decision makers on a local, regional, national and international level. EREC's membership includes the European federations for the different RE technologies, and seeks cooperation with RE organizations from around the world. It produces information materials on RES and encourages studies on the economics, techniques and feasibility in the field of renewable energy for the member associations and external organizations, as well as public and private institutions.

EREC's commercial focus considers the promotion of European exports of equipment and products, and the industrial protection of European RE technologies on global markets. It works to strengthen export initiatives and promote trade, and enlarge industrial export services with commercial publications, commercial indicators and seminars. See www.erec-renewables.org.

EREF (European Renewable Energies Federation) is a key European interest and lobby group for independent producers of electricity and fuels from all renewable energy sources (except large hydro). EREF members are national associations of RES producers and supporting agencies from many EU Member States. EREF represents at present approximately 17,000MW of installed capacity in RES electricity in the EU, and is representing a growing share of renewable energies in heating and cooling and in the transport sector. EREF provide detailed information for the public and media on renewables, as well as acting as a watchdog in the energy sector, and also as a promoter of regional sustainability. See: www.eref-europe.org

Eurosolar brings together expertise from the fields of politics, industry, science and culture to promote the introduction of solar energy; it promotes the total replacement of nuclear and fossil energies with RES; it regards a total solar energy supply as essential for a sustainable economy, and acts to change conventional political priorities and common infrastructures in favour of RE, from the local to the international level.

Eurosolar organizes key international conferences on renewables, including the instrumental 'renewables 2004' event. Eurosolar also initiated the World Council for Renewable Energies. See www.eurosolar.org.

The Feed-in Cooperation (FIC) is an organization set up between the BMU and the Spanish Ministry for Industry, Tourism and Trade, to help share experience on feed-in systems, to develop them and to assist other countries with implementation. The organization also hosts workshops for parties who have implemented, or wish to implement a feed-in law in their country, state or province. New members are welcome to join the Cooperation officially, but participation and information are available to all interested parties. The wealth of knowledge, resources and information is invaluable, and should be engaged with in order to learn from experience already gained in this highly contested area. The website provides access to many pertinent research papers and documents from FIC workshops, as well as country-specific legislation. See www.feed-in-cooperation.org.

A major proposal for a new body in the renewables field is IRENA – The International Renewable Energy Agency. Its main focus would be to increase the percentage of renewable energy in the global energy mix. Bringing together international experience and expertise in the field of renewables, IRENA would work with governments to help develop policies and institutional and technological capacity across the world, helping countries to develop their renewable generation capacities. It would further act as an advocate for renewables in the international political arena, in much the same way that the IAEA acts as a powerful international lobby for nuclear interests. IRENA would similarly promote the effective and efficient use of renewables. The IAEA began with a small number of partner countries, and IRENA would do the same, emerging initially from a coalition of proactive countries who prioritize ecological modernization, and in particular the renewable energy component of this.

ISEP in Japan is an independent, non-profit research organization, founded in 2000 by energy experts and climate change campaigners. Its activities include provision of resources and services to realize sustainable energy policies. The main areas of activity include promotion of RE, improvement of energy efficiency, and restructuring the energy market. It engages in a wide range of programmes and activities by providing policy recommendations with regard to RE and energy efficiency to the Japanese government, guidance and advice to local municipalities, and conducting events such as international conferences and symposia. ISEP also closely liaises with many organizations in Europe, the US and Asia to exchange information on sustainable energy policies. See www.isep.or.jp.

REN21 is a global policy network that provides a forum for international leadership on renewable energy. Its goal is to bolster policy development for the rapid expansion of RE in developing and industrialized economies. Open to a wide variety of dedicated stakeholders, REN21 connects governments, international institutions, NGOs, industry associations, and other partnerships and initiatives. Linking the energy, development and environment sectors, REN21 strengthens the influence of the unique RE community that came together at the 'renewables 2004' conference in Bonn. See www.ren21.net.

The IEA has set an Implementing Agreement on Renewable Energy Technology Deployment (RETD), in order to carry through selected activities that aim at accelerating the deployment of RE. This is done by targeting policy makers and private companies dealing with energy. The RETD also arose from the 'Renewables 2004' conference.

The aims of the RETD Implementing Agreement are to: improve cooperation between participating governments in identifying cross-cutting barriers to deployment and providing 'best practice' solutions; provide guidance to the private sector and policy makers on innovative business strategies and projects that encourage technology deployment by fostering public–private partnership projects; and to facilitate ongoing international dialogue and public awareness of RE deployment by contributing concrete examples of deployment solutions.

The general barriers to renewables such as administration, planning, legal, technical, grid connection and access, and costs and financing are main areas of work for the RETD. The replacement of fossil fuel heating and cooling systems with those driven by renewables is another key area for the organization. See www.iea-retd.org.

The WCRE (World Council for Renewable Energy) is a globally-operating independent non-commercial organization, which seeks to provide a clear and uncompromising voice for the worldwide replacement of fossil and nuclear energy by renewable energy. It challenges governments and international organizations to prioritize renewable energy deployment and rapidly replace conventional energies. It formulates political 'grand strategies' for renewable energy as opposed to the development of conventional energy markets.

The WCRE has, since its inception in 2001, organized many strategic international conferences on renewables such as the 2002 and 2004 World Renewable Energy Forums, the World Renewable Energy Assembly in 2005, two International Parliamentary Fora in 2004 and 2005 and the International Conference on the Storage of Renewable Energy. One of its main goals is the establishment of an International Renewable Energy Agency (IRENA). The WCRE works in partnership with other organizations such as EUROSOLAR and ACORE (American Council on Renewable Energy). See www.wcre.org.

12

The Future: Policy, Technology, Employment and Investment

The fossil fuel age is beginning to draw to a close, leaving a legacy that will be debated throughout the remainder of human history. It is bringing us perilously close to the limits of what nature can tolerate without huge change occurring, but swift action will give us the opportunity to keep our world habitable. Our future must be one powered by renewables, if we are to continue to lead lives in any way comparable to those we lead today. It is now time to speed the replacement of fossil fuels with renewable alternatives. It does indeed seem ridiculous that the status quo causes so much damage to life on Earth, yet the alternatives have long been the ones treated with suspicion. We must now grow beyond our adolescence as a species, and learn the meaning of boundaries, of limits, of responsibilities. Nature's systems demonstrate for us how living within one's limits, and wasting nothing, is the only way to survive over the long term. Harnessing free energy, then, is one of the first steps toward a more enlightened, secure way of life for our species.

This switch to using free, inexhaustible energy is an open invitation to all, yet it is the few who have so far taken up the offer, and begun to put in place a variety of policies and schemes that are making this future possible. This landscape of policy and technology is one of the most intensely researched and fastest-developing in the world. Technological breakthroughs are coming thick and fast, attracting increasing numbers of actors and ever-larger investment funding.

This chapter will briefly cover some recent advances in technology to provide an idea of how fast change is occurring, and in how many fields. The simple reason for doing so is that very few people are aware of the level of activity in the sector, and that such breakthroughs are occurring continually. Our views on renewables are almost always behind the times, and many developments are underway to bring down costs, increase production and improve efficiency.

The ideal platform for keeping up to date with this exciting field is now of course the web (www.renewableenergyaccess.com is a recommended English-language source for this). It will soon become clear from the following, whether one is familiar with the field or not, that renewables offer enormous positive opportunities for many issues and areas globally, including employment, transport, architecture, waste, rural and regional development, agriculture, forestry and assisting developing nations with clean energy.

SOLAR TECHNOLOGY

Nanosolar thin film technology

This technology may turn solar on his head if its claims are borne out in full production. Instead of producing silicon wafers, Nanosolar have patented a design for solar on a roll. A blend of metals including copper and gallium is placed on metal foil. The solar film is 100 times thinner (and has a clear advantage in materials costs) than a silicon wafer but is claimed to produce roughly the same amount of energy. The roll-printing production processes are said to be simple, robust, and an order of magnitude faster in throughput relative to vacuum-based thin-film production techniques. This greater efficiency should therefore greatly reduce labour, capital and process costs, and enable enormous production volume scalability. The combination of production factors should provide the most cost-effective solar technology yet. Investment in the project has been enormous, with significant funds coming from people behind eBay and Google. Production facilities are to be set up in San Jose, California (with an annual output of 400MW expected) and Berlin, Germany.

Mono2

The development of a new silicon growth process is claimed to increase the efficiency of multicrystalline-based cells, and will be able to produce 5 to 8 per cent more power than solar cells made with conventional processes, leading to a substantial cost reduction at the installed system level. The first modules to incorporate this process will be available in 2007 from BP Solar. The new product combines the silicon growth process with a new screen-printing process to improve solar cell and overall module efficiency. These Mono2 products will offer the efficiency and appearance of monocrystalline products at a lower production cost similar to multicrystalline products. The Mono2 casting process increases cell efficiency over traditional multicrystalline-based solar cells. It prints conductors onto the cells, allowing the cells to produce in excess of 2 per cent more power than those printed with conventional processes. These two new technologies combined result in an overall cell and module efficiency increase of roughly 8 per cent compared to their conventionally manufactured counterparts.

Concentrated solar thermal power

A collaborative project between the International Energy Agency (SolarPACES programme – Solar Power and Chemical Energy systems) and Solar de Almeria – a collaborative platform – will soon begin to feed solar electricity into Spain's electricity grid. The plant will run on Concentrated Solar Thermal Power (CSP), a technology that produces electric power by converting the sun's energy into high-temperature heat using various mirror configurations. The heat is then channelled through a conventional generator. It will be Europe's first commercial solar power plant of its sort, highlighting solar electricity's emergence as an increasingly viable and flexible source of energy.

The new advance is due in part to the evolution of existing technologies and in part to a ministerial ruling by the Spanish government in March 2004 which removed economic barriers to the grid-connection of RE. Spain has set an overall target for 29.4 per cent of its electricity to come from RES by 2010. By then, it predicts that some 500MW will be provided by CSP.

The trend is not confined to Spain. Greenpeace/ESTIA (European Solar Thermal Power Industry Association)/SolarPACES' scenarios, published in 2005, predict that solar thermal technology is destined to move from being a relatively modest RE source to a significant contributor in 2040 (meeting 5 per cent of the world's demand and avoiding some 50 million tonnes of CO_2 a year), alongside current market leaders like hydro and wind power (Mancini and Geyer, 2006).

SolarChill

SolarChill is a new ozone-layer friendly refrigeration technology running entirely on solar power, enabling the safe delivery of vaccines and food to regions of the world without electricity. SolarChill was born through separate discussions between Greenpeace, the United Nations Environment Programme (UNEP), and the World Health Organization (WHO). The challenge they addressed was the provision of affordable and environmentally safe refrigeration for the maintenance of vaccines and medicines, and the preservation of food, for parts of the world that have no electricity or have unreliable supplies of electricity. At present, refrigerators in developing countries usually use kerosene, propane and, to a far lesser extent, solar power. Medicine is most often stored in unreliable kerosene refrigerators.

SolarChill refrigerators do not require batteries or connection to the grid – they store power in three 60W voltaic panels. They do, however, have an AC/DC converter, which provides flexibility to use wind, hydropower, biogas or grid energy, or even a car battery when there is not enough sun. The coolers are also energy-efficient due to their excellent insulation. They do not contribute to global warming thanks to the special hydrocarbon system. Even maintaining a constant temperature, important where medicine is concerned, does not require an electronic controller – it is managed through natural convection methods. A single SolarChill unit can serve a population of 50,000 people for preserving vaccines, and will be 50–60 per cent cheaper than current solar refrigerator models.

Solar panel windows

A company by the name of Octillion has entered into a Sponsored Research Agreement with scientists at the University of Illinois for the development of a new patent-pending technology using nanosilicon PV solar cells that could convert normal home and office glass windows into ones capable of converting solar energy into electricity – without losing significant transparency or requiring major changes in manufacturing infrastructure, it is claimed.

Limited loss of transparency and minimal changes in manufacturing infrastructure are among the advantages envisioned for this solar innovation. The technological potential of adapting existing glass windows into ones capable of generating electricity from the solar energy has been made possible through the discovery of an electrochemical and ultrasound process that produces identically sized, 1–4 nanometres in diameter, highly luminescent nanoparticles of silicon that provide varying wavelengths of photoluminescence with high quantum down-conversion efficiency of short wavelengths. When thin films of silicon nanoparticles are deposited onto silicon substrates, ultraviolet light is absorbed and converted into electrical current. With appropriate connections, the film acts as nanosilicon PV solar cells that convert solar radiation to electrical energy. The advantages of this kind of breakthrough are clear, and being able to fit solar PV windows to new or existing buildings could be possible within the next few years.

Higher efficiency panels

Announced in December 2006, a PV cell produced by Spectrolab Inc, a wholly owned subsidiary of Boeing, and funded in part by the US Department of Energy (DOE), achieved a milestone in PV design with a conversion efficiency of 40.7 per cent. The breakthrough could lead to PV systems with an installed cost of $3/W and produce electricity at a cost of $0.08–0.10/kWh. The 40.7 per cent cell was developed using a structure called a multi-junction solar cell. This type of cell achieves a higher efficiency by capturing more of the solar spectrum. In a multi-junction cell, individual cells are made of layers, where each layer captures part of the sunlight passing through the cell – allowing the cell to absorb more energy from the sun's light. According to Spectrolab, high efficiency multi-junction cells have an advantage over conventional silicon cells in concentrator systems because fewer solar cells are required to achieve the same power output.

For the past two decades researchers have tried to break the 40 per cent efficient barrier on solar cell devices. In the early 1980s, DOE began researching what are known as 'multi-junction gallium arsenide-based solar cell devices' – multi-layered solar cells that converted about 16 per cent of the sun's available energy into electricity. In 1994, DOE's National Renewable Energy laboratory broke the 30 per cent barrier, which attracted interest from the space industry. Most satellites today use these multi-junction cells.

Almost all of today's solar cell modules do not concentrate sunlight but use only what the sun produces naturally, what researchers call 'one sun insolation', which achieves an efficiency of 12–18 per cent. However, by using an optical concentrator, sunlight intensity can be increased, squeezing more electricity out of a single solar cell. 'This solar cell performance is the highest efficiency level any photovoltaic device has ever achieved,' said Dr David Lillington, president of Spectrolab. 'The terrestrial cell we have developed uses the same technology base as our space-based cells. So, once qualified, they can be manufactured in very high volumes with minimal impact to production flow' (RenewableEnergyAccess.com, 2006a).

WIND

MagLev turbines

Chinese developers have designed the world's first permanent magnetic levitation wind power generator. The device is called a MagLev generator, and is being regarded as a key breakthrough in the evolution of global wind power technology. The generator was jointly developed by Guangzhou Energy Research Institute under China's Academy of Sciences. The MagLev generator is expected to boost wind energy generating capacity by as much as 20 per cent over traditional wind turbines. According to the chief scientist behind the technology, the generator can dramatically lower operational expenses of wind farms – by as much as 50 per cent. This, he claims, would drive the cost of wind power to below 5 cents per kWh. The MagLev is able to utilize winds with starting speeds as low as 1.5 metres per second, and cut-in speeds of 3 metres per second. The new technology could potentially fill the power void in locations with no connection to the grid by harnessing low-speed wind resources that were previously untappable. The technology is expected to create new opportunities in low-wind-speed areas worldwide such as mountain regions, islands, observatories and television transfer stations. In addition, the MagLev generator could provide roadside lighting along highways by utilizing the airflow generated from vehicles passing by (Thomas, 2006).

Stormrider turbine

A new design for turbines has come in the shape of the Stormrider turbine. Concept illustrations show a turbine that looks like a jet engine, housing all blades and moving parts within the structure. The advantages over traditional turbine designs claimed for it include:

- operation across double the current range of wind speeds (7mph to >120mph);
- double the average efficiency – it can convert up to 70 per cent of the wind power to electricity (apparently exceeding the theoretical Betz limit of 59 per cent);
- quieter and so less intrusive to people living nearby;
- non-threatening to birds and bats as it has no external blades;
- smaller and hence has less visual impact;
- more power produced per rotation;
- less maintenance required, and hence lower operating costs;
- operates at extreme wind speeds, so avoiding the need to shut down and be backed up by conventional energy plants.

If this design successfully comes to market, the wind industry could be revolutionized.

Floating turbines

In 2004, Massachusetts Institute of Technology (MIT) colleagues teamed up with wind turbine experts from the National Renewable Energy Laboratory (NREL) to integrate a wind turbine with a 'floater', or offshore floating platform. Their design calls for a tension leg platform (TLP), a system in which long steel cables, or 'tethers,' connect the corners of the platform to a concrete block or other mooring system on the ocean floor. The platform and turbine are thus supported not by an expensive tower but by buoyancy. According to their analyses, the floater-mounted turbines could work in water depths ranging from 30 to 200 metres. In the north-east of the US, for example, they could be 50 to 150 kilometres from shore. And the turbine atop each platform could be big – an economic advantage in the wind-farm business. The MIT–NREL design assumes a 5MW experimental turbine now being developed by industry. (Onshore units are generally 1.5MW; conventional offshore units, 3.6MW.)

Ocean assembly of the floating turbines would be prohibitively expensive because of their size. The wind tower is fully 90 metres tall, the rotors about 140 metres in diameter, so the researchers designed them to be assembled onshore, probably at a shipyard, and towed out to sea by a tugboat. To keep each platform stable, cylinders inside it are ballasted with concrete and water, and once on site the platform is hooked to previously installed tethers. Water is pumped out of the cylinders until the entire assembly lifts up in the water, pulling the tethers taut. It is estimated that building and installing this floating support system should cost a third as much as constructing the type of truss tower now planned for deep-water installations. Because of the strong offshore winds, the floating turbines should produce up to twice as much electricity per year (per installed MW) as wind turbines now in operation. In addition, since the turbines are not permanently attached to the ocean floor, they are a movable asset. If a company with a number of wind turbines serving one area needs more power for another, it could feasibly unhook some of the floating turbines and tow them to where they are needed (Stauffer, 2006).

WAVE ENERGY

Wave hubs

The world's first commercial wave energy project is to deliver wave-generated energy to the north of Portugal. The first stage of the EU-funded programme, the result of two decades of research at Lisbon's Superior Technical Institute, will bring the first 2.25MW ashore at Agucadoura, in northern Portugal, and will power 1500 homes through the national (state-run) electricity grid. Funded by a consortium headed by leading Portuguese RE company Enersis, the venture uses groundbreaking Pelamis wave devices manufactured by Edinburgh firm Ocean Power Delivery, considered the world's leading wave technology. The project is supported by state-run power company Energias de Portugal. With its geographical position and extensive coastline giving access to the larger and more powerful Atlantic waves, official estimates from Portugal's State Secretariat for Industry and Innovation have predicted wave power could account for up to 30 per cent of the country's GDP by 2050. RE experts have determined that wave farms in Portugal could yield as much as three times as much energy as that produced by a wind turbine park for the same investment cost.

The utilization of wave energy may have a significant socio-economic impact in Portugal, namely regarding renewables, job creation, opportunities to export equipment and services, innovation and development of technology, as well as companies dedicated to the exploitation of other oceanic resources. As part of the government supported alternative energy plan, another 28 wave power devices will be installed in Portugal within a year, reaching a target of 22.5MW of electricity produced using wave energy.

The Pelamis is a semi-submerged, articulated structure composed of cylindrical sections linked by hinged joints. The wave-induced motion of these joints is resisted by hydraulic rams, which pump high-pressure oil through hydraulic motors via smoothing accumulators. The hydraulic motors drive electrical generators to produce electricity. Power from all the joints is fed down a single umbilical cable to a junction on the sea bed. Several devices can be connected together and linked to shore through a single seabed cable.

Wave power is currently a largely untapped resource but is the most concentrated form of RE. In the UK for example, the total wave resource is equivalent to 2 to 3 times current electricity demand. However, it is not practical to recover all of this energy. The economically recoverable resource for the UK alone has been estimated to be 87TWh (terawatt hour) per year, or around 25 per cent of current UK demand. Just 5 per cent of this resource could provide a similar generation capacity to that of the nuclear industry. A 30MW wave farm would occupy a square kilometre of ocean and produce sufficient power for over 20,000 UK homes. All of the energy is concentrated near the water surface with little wave action below 50 metres depth. This makes wave power a highly concentrated energy source with much smaller hourly and day-to-day variations than other renewable resources such as wind or solar. Conveniently, the seasonal variation of wave power closely follows the trend for electricity consumption in Western Europe (Ocean Power Delivery, 2006).

BIOFUELS

'Monster Cane' biomass

'Monster Cane' is three metres tall and productive even in poor soil, holds up in droughts and typhoons, and it yields twice as many stems as most sugarcane. This new variety, named for its size as much as its vigour, is grown on a test field on the tiny island of Ie in Japan's southernmost prefecture of Okinawa. Researchers at Japanese beer maker Asahi Breweries Ltd predict that Okinawan farmers will be growing Monster Cane not only for sugar but also to fuel cars, raise cattle and fertilize farmland. Monster Cane is Japan's first variety designed to produce ethanol without sacrificing sugar output. It was jointly developed by Asahi and the National Agricultural Research Centre for Kyushu Okinawa Region, an administrative agency.

In the near future, the cane will be harvested to feed a pilot plant run by Asahi Breweries, which aims to test its technology for producing ethanol from cane at a cost of just ¥30 (25 cents) per litre, making it competitive with gasoline. Asahi aims to put its technology into practical application after completing tests at the pilot plant in 2010. Researchers also hope the new variety will breathe life into Japanese farming of sugarcane, an important part of crop rotation in Okinawa, by adding value to sugar production.

Fuel-use ethanol is currently not produced commercially in Japan as the country lacks the necessary excess farm produce and the costs involved are usually too high. Tokyo, however, has signed an international accord to cut emissions of CO_2 and other GHGs blamed for global warming, and is getting serious about promoting ethanol. Ethanol is carbon-neutral, as the CO_2 released in the combustion of the fuel is offset by the CO_2 captured by plants through photosynthesis. But its use raises the issue of balancing the supply of crops for use in ethanol production with supply of crops for food and feed. Critics also say ethanol is no solution to global warming if massive inputs of fossil fuels are required to grow the crops and power the facilities used to produce ethanol. To address these concerns, Asahi has developed a carbon-neutral process of producing ethanol from high-biomass sugarcane. Asahi says the new cane variety can produce three times as much ethanol as other strains, and slightly more sugar. It also yields more bagasse, or crushed sugarcane refuse, which is burnt to generate the electricity to run a sugar-ethanol plant.

Asahi estimates the yield of the new sugarcane at 37.4 tonnes per hectare excluding moisture, which can be processed into 7.1 tonnes of sugar, 4.3 kilolitres of ethanol and 24 tonnes of bagasse. This compares with the yield of a conventional cane type at 17.4 tonnes per hectare, sugar output at 6.9 tonnes, ethanol production at 1.4 kilolitres and bagasse volume at 7.8 tonnes, which is too small to produce sufficient energy for a processing plant. The volume of bagasse from high-biomass sugarcane is more than enough to generate energy for the Asahi plant. Surplus bagasse is used as bedding for premium beef cattle on Ie Island, and as fertilizer after being mixed with animal excrement. Ethanol produced at the Asahi plant is blended with gasoline to fuel cars owned by the Ie village office. Japan allows the production and sale of E3, a fuel that is 3 per cent ethanol and 97 per cent gasoline. An increasing number of Japanese farmers have been abandoning sugarcane production amid intensifying competition from cheap imported sugar and shrinking domestic sugar consumption (L'Express, 2006).

Biogas Energy Project

More than five million tonnes of food scraps go into California landfills each year, but a new demonstration project by the University of California, the Biogas Energy Project, will process eight tonnes of leftovers weekly from premier restaurants. The project's goal is to use organic matter such as food waste and yard clippings for generating energy, rather than rotting in landfills and producing more CO_2. Each ton of scraps can produce enough green electricity to power ten average California homes for one day. The Biogas Energy Project is the first large-scale demonstration in the US of a new technology developed in the past decade by Ruihong Zhang, a UC Davis (University of California) professor of biological and agricultural engineering. The technology, called an 'anaerobic phased solids digester', has been licensed from the university and adapted for commercial use. Zhang's system processes a wider variety of wastes than other anaerobic digesters, most of which are in use on municipal wastewater treatment plants and livestock farms. It works faster, turning waste into energy in half the time of other digesters.

Zhang's system produces two clean energy gases – hydrogen and methane. Other digesters produce only methane. The gases can be burned to produce electricity and heat, or to fuel cars, trucks and buses. Norcal Waste Systems of

San Francisco is supplying the waste for the project because it already collects restaurant leftovers for its composting operation near Vacaville. Every day, Norcal collects 300 tonnes of food scraps from 2000 restaurants in San Francisco and 150 more restaurants in Oakland. They calculated that energy can be harvested out of about half of the waste material that the state currently sends to landfills (ENS, 2006).

Methane to Markets initiative

Since the dawn of the industrial revolution, methane concentrations in the atmosphere have more than doubled, largely because of human activity. Methane now accounts for 16 per cent of all GHG emissions globally, with about 60 per cent of total methane emissions coming from anthropogenic sources. The remainder of methane emissions comes from natural sources: principally wetlands, gas hydrates, permafrost and termite digestion. Approximately 25 per cent of total methane emissions and 43 per cent of anthropogenic methane emissions come from the four anthropogenic sources targeted in the 'Methane to Markets' initiative:

- Agriculture (animal waste management) accounts for almost 2 per cent of total methane emissions and 4 per cent of anthropogenic methane emissions.
- Coal mining accounts for about 4 per cent of total methane emissions and 8 per cent of anthropogenic methane emissions.
- Landfills account for 8 per cent of total methane emissions and 13 per cent of anthropogenic methane emissions.
- Natural gas and oil systems account for 11 per cent of total methane emissions and for 16 per cent of anthropogenic methane emissions.

Coal mines, landfills, natural oil and gas systems, and agriculture (animal waste management) in the partner countries account for approximately 60 per cent of the total global anthropogenic methane emissions from these four sources, and 25 per cent of global methane emissions from all anthropogenic sources.

Methane is a GHG that is 23 times as effective at trapping heat in the atmosphere as CO_2. It also has a relatively short atmospheric lifetime of approximately 12 years. These two characteristics make methane emissions reductions particularly effective at mitigating global warming in the near term (i.e. the next 25 years). Fortunately, cost-effective technologies for capturing and using methane as a clean energy source are available in every sector on which the Methane to Markets partnership currently focuses. Methane capture and use projects can facilitate economic development and improve local living conditions. In addition, capturing methane from underground coal mines improves safety conditions by reducing explosion hazards (Methane to Markets.org, 2006).

Biofuel from burgers

The German biodiesel producer Petrotec is set to build a plant to produce 100,000 tonnes of biodiesel annually using waste edible oil. The company has signed a lease on land in the port of Emden and expects to complete the plant in 2007 at a cost of around €15 million. The company currently produces about 85,000 tonnes of biodiesel annually using waste edible oil which under EU rules can no longer be put into animal feed. Petrotec has extended a contract to import waste edible fats from the McDonald's fast food chain in France. In 2006 they imported around 5800 tonnes of waste fats from McDonald's French operations, and have also signed an agreement with an unnamed German fast food chain to collect waste fat from about 260 of the chain's restaurants (Planet Ark, 2006a).

ENERGY STORAGE

On the key subject of energy storage, previously considered a long-term barrier to renewables penetration, José Etcheverry writes:

> *Growing advances in a variety of fields such as information technology, forecasting and computer modelling, materials science, renewable energy generation, conservation and efficiency innovations are enabling decision-makers today to opt for sustainable energy options that have been advocated for decades by visionaries such as Amory Lovins, Hermann Scheer, and Preben Maegaard.*
>
> *Some of the key unifying concepts behind the sustainable energy revolution that is rapidly transforming the electricity and heating & cooling sectors are: increased reliance on distributed generation; multidirectional flows of information to constantly and accurately match changing energy needs with available generation options; greater co-location and smarter interconnection systems between generators and energy users; increased reliance on an interdependent combination of renewable energy, conservation and efficiency; and growing penetration of smart grid approaches and storage options.*
>
> *To capitalize on the aforementioned benefits, specialists in a variety of diverse settings are increasingly relying on smart grid techniques such as: advanced forecasting and modelling (e.g. to ensure that renewable resources such as wind and solar are utilized to their full potential); greater penetration of combined heat and power plants (e.g. to back-up and complement intermittent renewable sources); closely matching the operation of hydro-electric reservoirs with wind generation; and increased reliance on pumped-hydro options.*

Storage options, on the other hand, although essential to provide practical technical solutions to ensuring on demand heat availability, addressing intermittency of renewable energy sources, and enhancing power quality are less understood. Storage options are essential to: help avoid potential shut-downs (e.g. when power generation is not matched with consumption); to minimize the building of additional transmission infrastructure; to avoid investment in power plants for peak power needs; to make efficient use of combined heat and power systems; and to accumulate heat energy (e.g. from day to night or from summer to winter).

As well as a variety of advances in battery technology, the following table shows several non-battery storage systems currently at various stages of development, with some already relatively mature.

Table 12.1 *Non-battery storage systems*

Storage technology	Main options	Application examples	Status
Flywheels	Single and integrated-matrix High temp. super conductor (HTSC) magnetic bearings	Load balancing, frequency regulation and stabilization, urban traffic uses	Commercial demonstrations (e.g. California and NY)
Electric double-layer capacitors	Super/ultra capacitors Symmetric/asymmetric	Grid frequency stabilization Local reactive power support	Demonstration
Large-scale storage	Hydroelectric reservoirs	Peak load (PL) Reserve (R)	Commercial
Pumped hydro	Hydroelectric reservoirs	PL + R	Commercial
Compressed air (CAES)	a) Centralized options: storage in underground aquifers, caverns, heat recuperation, heat storage; b) decentralized (hydro-pneumatic)	Balance RE sources such as wind power Reduce need for additional grid capacity	a) Commercial demonstrations (e.g. Iowa Municipal Utilities) b) Demonstrations
Hydrogen	Electrolizers + fuel cells	Electrical and transportation	Commercial demonstrations

Source: Etcheverry, 2007

A variety of options and materials can currently be used for heat storage, for example: water, soil, concrete, phase change materials (medium and high temperature), zeolite, stationary or mobile storage systems, and thermo-chemical metal hydride.

Thermal storage is essential to greatly increase the widespread use of solar domestic hot water and combi-systems. Thermal storage *strategies are also essential to increase the efficiency of fossil-fuelled power generation (e.g. CHP) and to accumulate the significant flows of heat that are currently wasted in all kind of industrial applications (< 500 °C). Furthermore, storage is essential to support important renewable energy approaches such as solar thermal power plants.*

Increased research efforts in thermal storage are needed to concentrate on lowering costs and increase storage density for materials as well as systems to thereby exponentially increase the amounts of renewable energy used globally and also to greatly improve the efficiency of thermal generation systems.

As the previous sections indicate, achieving a global shift away from polluting sources of energy and toward renewable energy (RE) will require paying special attention to the mass development of storage solutions.

This imperative transition needs to be supported by a number of smart policies. A crucial global effort is required to phase-out and redirect towards the RE sector all the generous subsidies and market incentives that are currently dedicated to the fossil fuel and nuclear sectors.

This crucial step needs to be accompanied by an emphasis on prioritizing a wide variety of conservation and efficiency initiatives (e.g. real time electricity pricing, building and appliance codes) to reduce overall energy use and especially to minimize peak loads (e.g. a crucial step to reduce the use of expensive and polluting fossil fuel peaking plants).

In addition, international technological collaboration on renewable energy needs to increase especially in relation to greatly increased training opportunities at all educational levels.

At the regional and national level it is crucial to implement effective RE feed-in tariffs with particular attention on targeting premiums to ensure that they act as a market incentive to increase the capacity of renewable energy system operators to accurately forecast and dispatch renewable energy generation.

Since the long-term storage of physical energy is difficult, RE policies at the national and local level need to be targeted to facilitate profit maximization per storage cycle and to achieve increases in cycles per year; in addition, innovative RE policies need to be crafted to aim at minimizing operation, maintenance, and capital costs of storage systems (e.g. through the use of progressive tax strategies, accelerated capital depreciation strategies, and performance incentives). (Etcheverry, 2007)

LEGISLATION

European biofuels legislation

In an attempt to reduce GHGs in line with its Kyoto Protocol obligations, the EU adopted Directive/2003/30 EC, aimed at increasing the share of biofuels used in road transport from 0.8 per cent to 5.75 per cent by 2010. Biofuels not only produce fewer GHGs than oil and gas, but also are more abundant and domestically available. Furthermore, because, for the moment, agricultural crops (such as corn, sugar beet, palm oil and rapeseed) are the main source for biofuels (known as first-generation biofuels), the domestic production of biofuels could generate new income and employment opportunities for European farmers following the reform of the CAP.

Another type of biofuels (known as second-generation) can be obtained from ligno-cellulosic or 'woody' sources (such as straw, timber, woodchips or manure), but these fibre-rich materials can only be converted into liquid biofuels via advanced technical processes, many of which are still under development. The target set in 2003 is non-binding and most countries still have a lot of work to do in the next three and a half years if they wish to reach it. Nevertheless, increasing concern about securing long-term energy supplies, combined with rising oil prices and a number of fiscal and financial incentives, have led to a major increase in biofuel production in the EU-25, especially as regards the EU's biofuel of choice – biodiesel, which represents around 80 per cent of EU biofuel use. Biofuel production jumped from 1.9 million tonnes in 2004 to nearly 3.2 tonnes in 2005 – a 65 per cent annual increase. So far, the EU has mainly focused on using more first-generation biofuels, such as biodiesel and bioethanol. However, an increasing number of doubts are being raised about this strategy:

- Can first-generation biofuels really contribute to reducing GHG? In principle, biofuels are 'carbon-neutral', but some studies show that biofuels can actually produce more GHG than conventional fuels if one includes the emissions from agriculture, transport and processing involved in their production.
- Can first-generation biofuels compete with traditional fossil fuels? EU-produced biodiesel just breaks even at oil prices around €60 per barrel, while bioethanol only becomes competitive with oil prices of about €90 per barrel.
- Are first-generation biofuels driving up the price of food? Biodiesel production has significantly increased the consumption of rapeseed within the EU, driving the price of edible oils to record levels. Increased biofuel consumption is also likely to cause significant growth in the production and consumption of ethanol, pushing up the price of sugar.

- Are first-generation biofuels greener than traditional fossil fuels? In order to reach its 5.75 per cent biofuels target, Europe will have to rely on imports of ethanol from Brazil, where the Amazon is being burned to plant more sugar and soybeans, and Indonesia, where rainforest land is being cleared out to house palm oil plantations. Some environmental groups are already terming first-generation biofuels 'deforestation diesel' – the opposite of the environmentally friendly fuel it is supposed to be.

An increasing number of voices are therefore calling on the EU to focus its attention on 'second-generation' biofuels. The results of a public consultation carried out by the Commission between April and July 2006 in view of reviewing its biofuels strategy before the end of 2006 shows that the majority of stakeholders believe second-generation biofuels to be promising because:

- they have a more favourable GHG balance compared to most current biofuels;
- they can be produced at cost-competitive prices, especially if low-cost biomass is used;
- they are able to use a wider range of biomass feedstocks – they do not compete with food production;
- they offer a better fuel quality than first-generation biofuels.

The main pathway for second-generation bioenergy production in the EU is gasification – also called the biomass-to-liquid (BTL) pathway. It uses high temperatures, controlled levels of oxygen, and chemical catalysts to convert biomass into liquid fuels, including synthetic diesel and di-methyl ether (DME). Gasification generally requires large-sized facilities and big capital investments, which makes progress in this area slower than in others. Nevertheless, the BTL pathway can process lignin, which comprises about one-third of plant solid matter, and can thus achieve higher liquid yields, displacing more petroleum.

Finland announced in October 2006, when they held the EU presidency, that it intended to lead the way in terms of second-generation biofuels, notably by providing funding for new gasification equipment for VTT, the Technical Research Centre of Finland. The new equipment will allow synthesis gas to be refined from biomass for the production of diesel fuels. The gasification plant will be able to exploit any carbonaceous raw materials, for example, forest industry residues, bark, biomass from fields, refuse-derived fuels and peat. In its review of the biofuels directive, the Commission will be looking to enhance support to Member States for developing such second-generation technologies. It will also look into setting new targets for biofuels use and could even decide to make them mandatory. It is also likely to impose minimum environmental standards for biofuels production (Euractiv.com, 2006).

Recent feed-in law expansion

Several EU countries revised or supplemented their feed-in laws in 2005–2006, including Austria, the Czech Republic, France, Greece, Ireland, the Netherlands and Portugal. Austria supplemented its FITs with additional support of over €190 million ($240 million) in investment subsidies through 2012. The Czech Republic adopted a new feed-in law that establishes tariffs for all renewables technologies. France extended its feed-in law to cover re-powered and renovated facilities exceeding €800–1000 ($1000–1250 at exchange rate of mid-2006) per kW of new investment, which now qualify for higher tariffs. French tariffs are now a match for Germany's in some technologies, and signify a major development in this notably pro-nuclear country. Greece reduced permit requirements, set new tariffs, added solar thermal power, and provided subsidies and tax credits. Ireland replaced its competitive tendering system with a new feed-in policy and established new tariffs. The Netherlands revised FITs through to 2007. Portugal adopted a new tariff-calculation formula that accounts for technology, environmental impacts and inflation. And Italy's new national FIT for solar PV, established in 2004, became operational in 2005, with a first 100MW of allocations subscribed quickly and expectations for at least 60MW in 2006.

In 2006, Ontario became the second province in Canada, after Prince Edward Island, to enact an FIT. The Indian states of Karnataka, Uttaranchal and Uttar Pradesh also adopted FITs in 2005, bringing to six the number of Indian states with feed-in policies. The state of Maharashtra also updated its 2003 wind power feed-in policy to include biomass, bagasse and small hydropower generation. These new policies in Canada and India increased the numbers of states/provinces/countries worldwide with feed-in policies from 37 to 41. New production incentives, called limited FITs by some, were also appearing in several US states, including Minnesota, New Mexico and Wisconsin (REN21, 2006, p9).

Through personal communications, the policy grapevine suggests that feed-in systems are being explored in every corner of the globe at present, and at every level. Despite opposition from special interest groups, the policy continues to spread and develop as those implementing them learn from direct experience and from one another. Best practices are being drawn on by governments around the world, and the possibility of harmonization of EU RE support schemes with a feed-in system is being explored. Another interesting area of exploration is the hybrid system, where a feed-in system is crossed with a quota system. Expert opinion from policy researchers suggests that this will undo a lot of the good work of the best feed-in designs, and will again favour large utility companies at the expense of small investors and householders, but at the time of writing, firm proposals are yet to be published.

The future of the EEG

One of chief architects of the German feed-in law, Hermann Scheer, describes future steps for the law. He proposes providing a 'storage bonus' for on-site storage of electricity generated at RE installations. This would increase stability of supply, energy independence and cost transparency. It could trigger regional-ization of network systems by local and regional electricity producers. In addition, this would provide further flexibility for the producers and allow them to shape their own future at the end of the guaranteed period under the law.

Storage would increase predictability, allowing grids to be serviced more accurately, and peak power could be supplied more profitably. Facilities with integrated storage could also be introduced in countries with no laws mandating renewables introduction.

A further proposal is to divide the day into time periods, covering different levels of demand. Tariffs could then be adjusted to control supply during those periods, favouring production of solar for example, which would increase the market development of solar yet further. The off-peak periods would favour wind. This would be another step towards addressing the availability of these resources (Scheer, 2007).

INVESTMENT

Renewables and low-carbon technology attracted a record $100 billion in finance during 2006, according to New Energy Finance, a UK-based analyst. They stated that in three years development in the sector has occurred that they expected to take ten years. According to their figures, $70.9 billion of new investment came into the market in 2006, up 43 per cent on 2005. A further $29.5 billion came from mergers and acquisitions, leveraged buyouts and refinancing of assets. Some of the most substantial deals were done on the stock markets, with $10.3 billion raised via initial public offerings (IPOs), up from $4.3 billion in 2005 and $0.7 billion in 2004. Frankfurt saw the most IPO activ-ity, registering 15 clean energy deals, between them raising $2.6 billion. NASDAQ came in second, with 21 deals worth $1.7 billion, followed by the London Stock Exchange's AIM, with 29 deals worth $1.2 billion. Interest in public listings was spurred on in the first quarter of the year by the US President's state of the union address, in which he called for an end to the US 'addiction' to foreign oil. There was a second spurt after the publication of the UK government's Stern Review at the end of October, which warned of the financial consequences of failing to address climate change, and the US mid-term elections at the beginning of November, which saw control of Congress swing from the Republicans to the Democrats.

Solar and biofuels attracted the most investment on the public markets, with the technologies raising $4.4 billion and $2.5 billion respectively. Interest in solar nearly doubled, while the biofuels sectors raised ten times as much in 2006 as in 2005.

The volume of venture capital and private equity investment activity grew 167 per cent compared with 2005, to more than $7 billion. Most of this cash was invested in biofuels.

However, not every segment of the market had a prosperous 2006. Specialist carbon market service providers such as brokers and fund managers raised only $67 million from the public markets over the year, down from $465 million in 2005. New Energy Finance put this down to a combination of factors, including the crash of the price of carbon in the EU Emissions Trading Scheme in April (Environmental Finance, 2006).

In May 2006 Jefferies' Alternative Energy and Cleantech Conference hosted attendees from leading alternative energy companies, institutional investors, hedge funds and government agencies. More than 30 companies presented on alternative energy and clean technology topics, including speakers from DuPont and General Electric, making the conference one of the largest gatherings of cleantech and alternative energy professionals to date. Jeffrey H. Lipton, Managing Director of Cleantech Investment Banking at Jefferies, said that venture-capital investment in cleantech was at $730 million in 2001, but should reach approximately $2 billion in 2006. Further, more and more companies are receiving venture-capital funding every quarter, with the average deal size at approximately $6.6 million in 2005, compared with $6.0 million in 2002. A real and sustainable rapidly growing sector is emerging in the area of alternative energy and clean technology.

Investment by North American venture capitalists in renewable fuels and other so-called clean technologies hit a record $933 million in the third quarter of 2006, according to industry group Cleantech Venture Network LLC. It was the ninth consecutive quarter of growth, a 10.8 per cent rise over the previous quarter, and a 120 per cent increase over the same quarter last year, according to researchers at the Michigan-based group. Year-to-date clean technology investments totalled $2.29 billion, double the $1.1 billion invested in the first three quarters last year, Cleantech said. The main clean technology investments were in energy. About $512 million was invested in biofuels, such as ethanol and biodiesel, while $69 million was invested in solar power. This investment is considered to be a result of the very high cost of oil and US states establishing mandates on biofuels. Many Silicon Valley venture capitalists, such as Vinod Khosla of Sun Microsystems, are funding start-ups in ethanol, which they see as a low-emission alternative to America's reliance on petroleum (Planet Ark, 2006b).

By 2006, at least 85 publicly traded RE companies worldwide (or RE divisions of major companies) had a market capitalization greater than $40 million. The number of companies in this category expanded significantly, from around 60 in 2005. The estimated total market capitalization of these companies and divisions in mid-2006 was more than $50 billion. Major additions during 2005–2006 included Suntech Power (China), Suzlon (India), REC

(Norway) and Q-cells (Germany), all with high-profile initial public offerings (IPOs).

The largest number of companies is in the solar PV industry, which is becoming one of the world's fastest growing, most profitable industries. Global production increased from 1150MW in 2004 to over 1700MW in 2005. Japan was the leader in cell production (830MW), followed by Europe (470MW), China (200MW) and the US (150MW). As in recent years, shortages of silicon continued to affect production. Capacity expansion plans by the solar PV industry for 2006–2008 total at least several hundred MW and potentially two GW.

RE market growth is happening in almost all sectors, countries and investment stages. In a few sectors the industry is facing production bottlenecks, but when these open up, renewed growth will be seen. Interest in renewables investment has spiked in the last two years in particular. New Energy Finance, an investment consultancy, estimates global financing flows for the clean energy sector at $59 billion in 2005, including mergers and acquisitions ($15 billion), asset finance for large-scale projects ($18 billion), 'distributed projects' or small-scale renewable investments ($7 billion), government R&D ($6 billion), corporate R&D ($4 billion), corporate plant and equipment ($3 billion), public equities investment ($4.3 billion) and venture capital for plant and equipment ($1.6 billion). Overall, the clean energy sector has gone well beyond idealists to include serious entrepreneurs. Investors are feeling more and more comfortable committing capital for the long periods required by most projects in this sector, particularly as technologies mature, expertise increases, and risk management is better understood.

An overview of RE finance and investment reveals ten key trends:

1 Global investment is growing in all regions, but emerging markets are expected to become the core.
2 Sector preferences reflect the maturity and potential of the technology.
3 Private investors are becoming more bullish due to policies and political factors.
4 Investment banking firms are showing interest.
5 Venture capital favours clean energy.
6 The bond market is starting to finance wind farms.
7 Private equity investments slowed in 2006.
8 RE company valuations have skyrocketed.
9 Mergers and acquisitions remain buoyant.
10 Market growth looks set to remain sturdy. (REN21, 2006)

Cost-competitiveness

More than half of participants surveyed at the Jefferies Alternative Energy and Cleantech Conference 2006 believe solar power will be cost-competitive with grid-generated electricity by 2015, while only one-quarter said solar would be competitive by 2010. Jefferies & Company released these alternative energy-related findings from the conference held in May 2006 in New York City. The supply of silicon continues to be the largest barrier to growth, and that trend is expected to continue for the next couple of years. However, current incentive structures will drive production and economies of scale, which will ultimately lower costs. In addition, the emerging raft of new technology and improved efficiency will lower costs further (Jefferies, 2006, p4).

Solar power systems today are more than 60 per cent cheaper than 1990. The theory of the learning curve shows that every doubling of PV output leads to a 20 per cent fall in price. This has also been confirmed in Germany: since 1990 the price of PV systems has fallen over 60 per cent from €13,500 to about €5000 today. Between 1999 and 2003, the fall in price was 25 per cent in the 100,000 roofs scheme. By way of international comparison, prices of solar power modules show a continual downward trend. However, further price decreases are only possible if mass production undergoes further expansion, with thorough R&D in progress at the same time (Renewableenergyaccess.com, 2006b).

The German RES act envisages a reduction of 5–6.5 per cent per annum in refunds for solar power fed into the grid. The average price of 1kWh of solar power will decrease nominally at 5 per cent per annum from 49 cents today to 23 cents in 2020. Conventional power on the other hand will become dearer. At a minor increase of 2.5 per cent per annum, the price of power will rise for the private consumer from 19.6 cents/kWh today to 28 cents/kWh in 2020. This way, solar power for the private customer will be cheaper from 2018 than obtaining conventional power (Friedrich Ebert Foundation, 2006).

JOBS

While job creation in RE in Europe has been very healthy, the US has yet to truly explore the potential. According to a study by the UCS, more than 355,000 jobs would be created if the US obtained 20 per cent of its electricity from wind, solar and other RES. This total is nearly double the number of jobs from generating the same amount of electricity from fossil fuels. In such circumstances, consumers would also save more than $35 billion on their electricity bills through to 2020, and another $14 billion in lower natural gas bills. Renewables achieve these savings primarily by reducing the demand for, and the price of, natural gas. The analysis found 20 per cent renewable electricity by 2020 would boost the US economy with benefits such as: a net gain of more than 157,000 new jobs in manufacturing, construction, operation, maintenance and other industries; $73 billion in capital investment; $16 billion in income to farmers, ranchers and rural landowners for biomass energy supplies and wind power land leases; and $5 billion in property tax revenues for rural communities (UCS, 2005a).

129

Instead of losing almost 80,000 jobs from chemical companies moving plants overseas to escape high natural gas prices, the US and other industrialized countries could be creating highly skilled RE jobs. Installing and operating wind turbines and solar panels and growing energy crops are jobs that cannot be outsourced. US power plant CO_2 emissions – a major contributor to global warming – would be 15 per cent lower with such a shift to RE.

As mentioned in Chapter 4, Germany expects the renewables sector to employ 500,000 people by 2020. Spain predicts that by 2010 their figure will be around 100,000. To assess this elsewhere, more research must be carried out on the potential for RE jobs in different countries in the developed and developing world. China will almost certainly be making headlines over the next few years with regard to how many people find employment in all areas of the renewables sector, particularly in the manufacture and export of cheap solar panels and wind turbines. While this will bring down costs for end users, it will undermine domestic production opportunities elsewhere, and should be taken as another signal to accelerate engagement with the industry, especially where suitable expertise, innovative capacity and infrastructure already exist. The workshop of the world that China has become will not wait for countries like the US to get its act together. Narrow, vested interests are delaying the development of renewables in every country one examines, yet the potential for improving the economic fortunes and energy security of industrialized nations in particular is only being exploited by a few countries, with Germany again taking a lead in this respect.

CONCLUSION

The material in this chapter is simply a snapshot of the latest advances in RE technologies, numbers of jobs being created around the world, investments being made, new feed-in policies being implemented or developed and so on. By the time you read this, matters will have improved further. The renewables sector is one of the most dynamic, and has perhaps the greatest potential of any to deliver a new global industry with vast peripheral benefits – simply through the array of vital needs that RE addresses, especially at this make or break time in our collective history. Until the true value of renewables is recognized, the barriers are removed, the myths are finally put to rest and a massive global acceleration of renewables deployment takes place, we will continue to gallop in the wrong direction. This is demonstrated by events around the world every day, and is being recognized at the highest levels. We have the technology, the policy instruments and the lessons to take us in the right direction – we simply need the will to do what is urgently called for.

Recommended Reading

Beck, F. and Martinot, E. (2004) 'Renewable energy policies and barriers', in C. J. Cleveland (ed) *Encyclopaedia of Energy*, Academic Press/Elsevier Science, London, San Diego, pp365–383

BMU (Federal Ministry for the Environment, Nature Conservation and Nuclear Safety) (2004) 'The main features of the Act on granting priority to renewable energy sources (Renewable Energy Sources Act) of 21 July 2004', available at http://erneuerbare-energien.de/files/english/renewable_energy/downloads/application/pdf/eeg_gesetz_merkmale_en.pdf

BMU (2005) 'The VDEW proposal for a so-called "integrative model" for the support of renewable energies in the electricity sector', information paper, available at www.bmu.de/english/renewable_energy/current/doc/36309.php

BMU (2006b) 'Renewable energy: Employment effects – impact of the expansion of renewable energy on the German labour market', available at www.bmu.de/files/pdfs/allgemein/application/pdf/employment_effects_061211.pdf

Bradford, T. (2006) *Solar Revolution*, The MIT Press, Cambridge, MA

Bundestag (2004) 'Act revising the legislation on renewable energy sources in the electricity sector', available at www.eurosolar.org/new/en/downloads/eeg_en.pdf

Butler, L. and Neuhoff, K. (2004) 'Comparison of feed in tariff, quota and auction mechanisms to support wind power development', Cambridge Working Papers in Economics CWPE 0503, CMI Working Paper 70, available at www.electricitypolicy.org.uk/pubs/wp/ep70.pdf

Carbon Trust and LEK Consulting (2006) 'Policy frameworks for renewables – analysis on policy frameworks to drive future investment in near and long-term renewable power in the UK', available at www.cleanenergystates.org/international/downloads/Policy_Frameworks_for_Renewables_Carbon_Trust_July2006.pdf

Coenraads, R., Voogt, M. and Morotz, A. (2006) 'Analysis of barriers for the development of electricity generation from renewable energy sources in the EU-25', ECOFYS, available at www.optres.fhg.de/results/OPTRES_D8_barriers.pdf

European Wind Energy Association (2005) 'Support schemes for renewable energy – A comparative analysis of payment mechanisms in the EU', available at www.ewea.org/fileadmin/ewea_documents/documents/projects/rexpansion/050620_ewea_report.pdf

Gipe, P. (2006) 'Renewable energy policy mechanisms', available at www.windworks.org/FeedLaws/RenewableEnergyPolicyMechanismsbyPaulGipe.pdf

Greenpeace and Global Wind Energy Council (2005) 'Wind Force 12: A blueprint to achieve 12 per cent of the world's electricity from wind power by 2020', available at www.ewea.org/fileadmin/ewea_documents/documents/publications/reports/wf12-2005.pdf

IDAE (Instituto para la Diversificación y el Ahorro de la Energía) (2005) 'Summary of the Spanish Renewable Energy Plan 2005–2010', available at www.bv-pv.at/upload/536_Summary%20Spanish%20Energy%20Strategy%202005-2010.pdf

Kissel, J. and Oeliger, D. (2004) 'Ein dreistes schelmenstück, Das stromeinspeisegesetz als einfallstor für die markteinführung von erneuerbaren energien', *Solarzeitalter*, vol 1, pp12–19.

Klein, A., Held, A., Ragwitz, M., Resch, G. and Faber, T. (2006) 'Evaluation of different feed-in tariff design options', Best Practice paper for the International Feed-in Cooperation, available at www.feed-in-cooperation.org/images/files/best_practice_paper_final.pdf

Lauber, V. and Mez, L. (2004) 'Three decades of renewable electricity policies in Germany', *Energy and Environment*, vol 15, no 4, pp599–623, available at www.ontario-sea.org/ARTs/Germany/Three%20decades%20of%20renewable%20electricity%20policy%20in%20Germany.doc

Lewis, J. and Wiser, R. (2005) 'Fostering a renewable energy technology industry: An international comparison of wind industry policy support mechanisms', LBNL-59116, November 2005, available at http://eetd.lbl.gov/ea/ems/reports/59116.pdf

Martinot, E., Wiser, R. and Hamrin, J. (2005) 'Renewable energy policies and markets in the United States', available at www.resource-solutions.org/lib/librarypdfs/IntPolicy-RE.policies.markets.US.pdf, accessed 20 January 2007

Mitchell, C., Bauknecht, D. and Connora P. (2003a) 'Effectiveness through risk reduction: A comparison of the renewable obligation in England and Wales and the feed-in system in Germany', *Energy Policy*, vol 34, no 3, pp297–305

Mitchell, C., Bauknecht, D. and Connora P. (2003b) 'Risk, innovation and market rules: A comparison of the renewable obligation in England and Wales and the feed-in system in Germany', WBS Centre for Management under Regulation (CMUR) Discussion Paper, available at http://users.wbs.ac.uk/group/cmur

Morris, C. (2006) *Energy Switch: Proven Solutions for a Renewable Future*, New Society Publishers, Gabriola Island, Canada

Planet Ark (2006) 'North American clean technology spending hits record', available at www.planetark.com/dailynewsstory.cfm/newsid/38654/story.htm

Ragwitz, M. and Huber, C. (2004) 'Feed-in systems in Germany and Spain: A comparison', Fraunhofer Institute: Systems and Innovation Research, available at www.erneuerbare-energien.de/files/english/renewable_energy/downloads/application/pdf/langfassung_einspeisesysteme_en.pdf

Ragwitz, M., Resch, G., Faber, T. and Huber, C. (2005) 'Monitoring and evaluation of policy instruments to support renewable electricity in EU member states', Fraunhofer Institute: Systems and Innovation Research, available at www.bmu.de/files/erneuerbare_energien/downloads/application/pdf/isi_zwischenbericht.pdf

REN21 (2006) 'Renewables Global Status Report – 2006 Update', available at www.ren21.net/globalstatusreport/download/RE_GSR_2006_Update.pdf

Sawin, J. (2004) 'National policy instruments: Policy lessons for the advancement and diffusion of renewable energy technologies around the world', Thematic Background Paper, Worldwatch Institute, Washington DC

Scheer, H. (2005b) *The Solar Economy: Renewable Energy for a Sustainable Energy Future*, Earthscan, London

Scheer, H. (2007) *Energy Autonomy: The Economic, Social and Technological Case for Renewables*, Earthscan, London

Teske, S. et al, Greenpeace International and European Renewable Energy Council (EREC) (2007) 'Energy [r]evolution: A sustainable world energy outlook', Greenpeace International, EREC, available at http://news.bbc.co.uk/nol/shared/bsp/hi/pdfs/25_01_07_energy_revolution_report.pdf

Toke, D. and Marsh, D. (2006) 'Will planners tilt towards windmills?', Report of Sustainable Technologies Programme Research Project into planning and financial issues surrounding wind power, Economic and Social Research Council, London available at www.sustainabletechnologies.ac.uk/final%20pdf/Project%20Innovation%20Briefs/Innovation%20Brief%201.pdf

Bibliography

Aulich, H. A. (2006) 'Opportunities for PV – Applications and market deployment in developing countries', presentation at the first General Assembly of the European Photovoltaic Technology Platform, 19 May, Brussels

Bechberger, M. and Reiche, D. (2005a) 'Europe banks on fixed tariffs', *New Energy*, no 2, April, pp14–17

Bechberger, M. and Reiche, D. (2005b) 'The spread of renewable energy feed-in tariffs (REFITs) in the EU-25', Berlin Conference 2004, 'Greening of Policies Interlinkages and Policy Integration', 3–4 December, Berlin

Bechberger, M. and Reiche, D. (2006) 'Good environmental governance for renewable energies – The example of Germany – Lessons for China?' WZB-Discussion paper no 6/2006, Berlin, available at http://skylla.wzberlin.de/pdf/2006/p06-006.pdf

Beck, F. and Martinot, E. (2004) 'Renewable energy policies and barriers', in C. J. Cleveland (ed) *Encyclopaedia of Energy*, Academic Press/Elsevier Science, London, San Diego, pp365–383

Bhattacharyya, S. C. (2006) 'Energy access problem of the poor in India: Is rural electrification a remedy?', *Energy Policy*, vol 34, pp3387–3397

Bird, L., Bolinger, M., Gagliano, T., Wiser, R., Brown, M., Parsons, B. (2005) 'Policies and market factors driving wind power development in the United States', *Energy Policy*, vol 33, no 11, pp1397–1407

Blanco, I. (2005) 'The way forward: Experiences with the feed-in system in Spain', presentation, IDAE (Instituto para la Diversificación y Ahorro de la Energía), December, Berlin

BMU (Federal Ministry for the Environment, Nature Conservation and Nuclear Safety) (2000) Nationales Klimaschutzprogramm – Beschluss der Bundesregierung vom 18 Oktober 2000, Berlin

BMU (2004) 'The main features of the Act on granting priority to renewable energy sources (Renewable Energy Sources Act) of 21 July 2004', available at http:// erneuerbare-energien.de/files/english/renewable_energy/downloads/application/pdf/ eeg_gesetz_merkmale_en.pdf

BMU (2005) 'The VDEW proposal for a so-called "integrative model" for the support of renewable energies in the electricity sector', information paper, available at www.bmu.de/english/renewable_energy/current/doc/36309.php

BMU (2006a) 'Renewable energies: Innovations for the future', available at www.bmu.de/ files/english/renewable_energy/downloads/application/pdf/broschuere_ee_ innovation_eng.pdf

BMU (2006b) 'Renewable energy: Employment effects – impact of the expansion of renewable energy on the German labour market', available at www.bmu.de/files/ pdfs/allgemein/application/pdf/employment_effects_061211.pdf

BMU (2006c) 'Trends in renewable energies in 2005 – current situation', available at www.bmu.de/files/erneuerbare_energien/downloads/application/pdf/ee_aktueller-sachstand_en.pdf

BMU (2006d) 'Environmental Policy: Renewable energy sources in figures – national and international development', available at www.bmu.de/files/english/renewable_energy/downloads/application/pdf/broschuere_ee_zahlen_en.pdf

Bode, S. (2006) 'On the impact of renewable energy support schemes on power prices' HWWI Research Papers, by the HWWI Research Programme 'International Climate Policy', September 2006, Hamburg, available at www.hwwi.org/fileadmin/hwwi/Publikationen/Research/Paper/Klimapolitik/HWWI_Research_Paper_4-7.pdf

Bradford, T. (2006) *Solar Revolution*, The MIT Press, Cambridge, MA

Bundestag (2004) 'Act revising the legislation on renewable energy sources in the electricity sector', available at www.eurosolar.org/new/en/downloads/eeg_en.pdf

Busch, P., Jörgens, H. and Tews, K. (2005) 'The global diffusion of regulatory instruments: The making of a new international environmental regime', *Annals*, American Academy of Political & Social Science, vol 598, no 1, pp146–167

Bustos, M. (2004) 'The new payment mechanism of RES-E in Spain', introductory report, Spanish Renewable Energy Association

Butler, L. and Neuhoff, K. (2004) 'Comparison of feed in tariff, quota and auction mechanisms to support wind power development', Cambridge Working Papers in Economics CWPE 0503, CMI Working Paper 70, available at: www.electricitypolicy.org.uk/pubs/wp/ep70.pdf

BWE (German Wind Energy Association) (2006) 'Energy Economy', available at www.wind-energie.de/en/wind-energy-in-germany/energy-economy/

Carbon Trust and LEK Consulting (2006) 'Policy frameworks for renewables – analysis on policy frameworks to drive future investment in near and long-term renewable power in the UK', available at www.cleanenergystates.org/international/downloads/Policy_Frameworks_for_Renewables_Carbon_Trust_July2006.pdf

Coenraads, R., Voogt, M. and Morotz, A. (2006) 'Analysis of barriers for the development of electricity generation from renewable energy sources in the EU-25', ECOFYS, available at www.optres.fhg.de/results/OPTRES_D8_barriers.pdf

Credit Suisse (2006) 'China's renewable energy sector: China's next booming sector', available at www.frankhaugwitz.info/doks/general/2006_07_China_RE_Credit_Suisse_Chinese_renewable_energy_sector_report.pdf, accessed 26 January 2007

Deepchand, K. (2002) 'Promoting equity in large-scale renewable energy development: The case of Mauritius', *Energy Policy*, vol 30, pp1129–1142

DSIRE (Database of State Incentives for Renewables and Efficiency) (2006a) 'California incentives for renewable energy, emerging renewables rebate program', available at www.dsireusa.org/library/includes/incentive2.cfm?Incentive_Code=CA30F&state=CA&CurrentPageID=1&RE=1&EE=1, accessed 18 December 2006

DSIRE (2006b) 'California incentives for renewable energy, public benefits funds for renewables & efficiency', available at www.dsireusa.org/library/includes/incentive2.cfm?Incentive_Code=CA05R&state=CA&CurrentPageID=1&RE=1&EE=0, accessed 18 December 2006

DSIRE (2006c) 'California incentives for renewable energy, renewables portfolio standard', available at www.dsireusa.org/library/includes/incentive2.cfm?Incentive_Code=CA25R&state=CA&CurrentPageID=1, accessed 18 December 2006

Economist (2005) 'China's courts: Winning is only half the battle', article, March

EIA (Energy Information Administration) (2006) 'Country analysis briefs: China', available at www.eia.doe.gov/emeu/cabs/China/Electricity.html, accessed on 26 January 2007

Ekostaden (2005) 'Västra Hamnen the Bo01-area: A city for people and the environment', available at www.ekostaden.com/pdf/vhfolder_malmostad_0308_eng.pdf, accessed 26 January 2007

Elliot, D. (2003) *A Solar World Climate Change and the Green Energy Revolution*, (Schumacher Briefings no10), Green Books, Dartington

Ender, C. (2006) 'Wind energy use in Germany – status 30.06.2006', *DEWI Magazine*, no 29, August

ENS (Environmental News Service) (2006) 'New technology turns bay area table scraps into fuels', available at www.ens-newswire.com/ens/oct2006/2006-10-24-09.asp, accessed 24 October 2006

Environmental Finance (2006) 'Clean energy financing reaches $100bn in 2006', www.environmental-finance.com/owcinews/0721nef.htm

Etcheverry, J. (2007) 'Developing renewable energy to their full potential through smart grids and storage options', pre-publication draft, David Suzuki Foundation, Vancouver

Euractiv.com. (2006) 'Can biofuels cure our oil dependency?', www.euractiv.com/en/environment/biofuels-cure-oil-dependency/article-159043, accessed 24 October 2006

European Commission (2004) Communication from the Commission to the Council and the European Parliament: 'The share of renewable energy in the EU', Commission Report in accordance with Article 3 of Directive 2001/77/EC, evaluation of the effect of legislative instruments and other Community policies on the development of the contribution of renewable energy sources in the EU and proposals for concrete actions {SEC(2004) 547}

European Commission (2005) Communication from the Commission: 'The support of electricity from renewables energy sources' {SEC(2005) 1571}

European Commission Directorate-General for Energy and Transport (2005) In *Figures: Energy & Transport 2005 Part 2: Energy*, available at http://ec.europa.eu/dgs/energy_transport/figures/index_en.htm

European Parliament (2001) 'Directive 2001/77/EC of the European Parliament and of the Council of 27 September 2001 on the promotion of electricity produced from renewable energy sources in the internal electricity market', *Official Journal of the European Communities*

European Photovoltaic Industry Association (2005) 'Position paper on a feed-in tariff for photovoltaic solar electricity', available at www.epia.org/documents/FeedInTariffEPIA.pdf

European Renewable Energy Council (2005) 'Position paper on the future of support systems for the promotion of electricity from renewable energy sources', available at www.erec-renewables.org/documents/Position_Paper/Electricity/PosPap_RES-E_final_201204.pdf

European Wind Energy Association (2005) 'Support schemes for renewable energy – A comparative analysis of payment mechanisms in the EU', available at www.ewea.org/fileadmin/ewea_documents/documents/projects/rexpansion/050620_ewea_report.pdf

Eurosolar (2005a) 'Eurosolar Awards Appreciations', available at: www.eurosolar.org/new/en/esp_2005.html

Eurosolar (2005b) 'Five reasons for a feed-in model, five reasons against quota systems', Eurosolar factsheet

Eurosolar (2005c) 'The success of feed-in tariffs in Europe', Eurosolar factsheet

Eurosolar Amendment to the Renewable Energy Sources Act (EEG) (2004) available at www.eurosolar.org/new/en/downloads/eeg_begruendung_en.pdf

Feed-in Cooperation (2005) 'Summary of main results International Feed-in Cooperation (FIC) Second workshop', Berlin 15–16 December, available at www.feed-in-cooperation.org/images/files/summary_workshop_fic.pdf

Friedman, T. (2006) *The World is Flat: The Globalized World in the Twenty-first Century*, Penguin, London

Friedrich Ebert Foundation (2006) 'Solar energy in Germany: Focus on Germany', available at http://library.fes.de/pdf-files/bueros/london/03560.pdf, accessed 25 January 2007

German Wind Energy Association (2006) 'Energy economy', available at www.wind-energie.de/en/wind-energy-in-germany/energy-economy/, accessed September 2006

Gipe, P. (1995) *Wind Energy Comes of Age*, John Wiley and Sons Ltd, New York

Gipe, P. (2006) 'Renewable energy policy mechanisms', available at www.wind-works.org/FeedLaws/RenewableEnergyPolicyMechanismsbyPaulGipe.pdf

Girardet, H. (2004) *Cities, People, Planet*, John Wiley and Sons Ltd, Chichester

Global Wind Energy Council (2005) 'Record year for wind energy: Global wind power market increased by 43 per cent in 2005', press release

Grameen Communications (2006) 'Grameen Shakti: Development of renewable energy resources for poverty alleviation', available at www.grameen-info.org/grameen/gshakti/index.html, accessed on 26 January, 2007

Greenpeace (2006a) 'Decentralising energy – the Woking case study', available at www.greenpeace.org.uk/MultimediaFiles/Live/FullReport/7468.pdf, accessed 26 January 2007

Greenpeace and European Photovoltaic Industry Association (2006b) 'Solar generation – Solar electricity for over one billion people and two million jobs by 2020', available at www.greenpeace.org/raw/content/international/press/reports/solargen3.pdf

Greenpeace and Global Wind Energy Council (2005) 'Wind Force 12: A blueprint to achieve 12 per cent of the world's electricity from wind power by 2020', available at www.ewea.org/fileadmin/ewea_documents/documents/publications/reports/wf12-2005.pdf

Haas, R., Resch, G., Ragwitz, M. and Faber, T. (2006) 'Optimal promotion strategies for increasing the share of RES-e – lessons from the OPTRES project', presentation at Maribor, 10 May 2006, available at www.ljudmila.org/sef/stara/srecanje100506_gradiva/Prvi_dan/Hass.pdf

Head, J. (2006) 'Japan races to hit Kyoto targets', available at http://news.bbc.co.uk/1/hi/sci/tech/4720978.stm, accessed 20 January 2006

Hemke, S. (2006) 'The renewable energy sources act & the feed-in cooperation', presentation, Maribor, 10 May 2006, available at www.ljudmila.org/sef/stara/srecanje100506_gradiva/Prvi_dan/Hemke.pdf

Herring, H. (2006) 'Confronting Jevons' Paradox: Does promoting energy efficiency save energy?' International Association for Energy Economics, available at www.econ.surrey.ac.uk/staff/lhunt/Teaching/EC456/Herring%20(2006).pdf, accessed 26 January 2007

IDAE (Instituto para la Diversificación y el Ahorro de la Energía) (2005) 'Summary of the Spanish Renewable Energy Plan 2005–2010', available at www.bv-pv.at/upload/536_Summary%20SPanish%20Energy%20Strategy%202005-2010.pdf

IEA (International Energy Agency) (2003a) *Cool Appliances: Policy Strategies for Energy Efficient Homes*, OECD/IEA, France

IEA (2003b) 'National Survey Report of PV power applications in Japan 2002', May, Japan, available at www.iea-pvps.org/nsr02/download/jpn.pdf

IEA (2004) *Coming in From the Cold: Improving District Heating in Transition Economies*, OECD/IEA, France

IEA (2005) 'Standby power use and the IEA "1-Watt Plan"', factsheet, OECD/IEA, France

IEA (2006) *Barriers to Technology Diffusion: The Case of Compact Fluorescent Lamps*, OECD/IEA, France

Iida, T. (2001) 'Greening Energy Policy in Japan – green energy movement and the changes in renewable energy policy', unpublished

Jefferies (2006) 'Solar energy expected to be cost-competitive in less than 10 years according to a majority at conference participants polled', available at www.jefferies.com/cositemgr.pl/htm/OurFirm/NewsRoom/PressReleases/2006/20060618press.shtml

Jones, J. (2005) 'Growth without subsidy?' available at www.earthscan.co.uk/news/article/mps/UAN/350/v/3/sp/332252698778344281270#jackiejones, accessed 24 January 2007

Junfeng, L., Jinli, S. and Lingjuan, M. (2006) 'China: Prospect for renewable energy development', available at www.hm-treasury.gov.uk/media/999/B2/Final_Draft_China_Mitigation_Renewables_Sector_Research.pdf, accessed 26 January 2007

Karekezi, S., Kithyoma, W. and Muzee, K. (2007) 'Bagasse-based cogeneration in Mauritius – successful energy policy interventions in Africa', AFREPREN/FWD, Nairobi

Kissel, J. and Oeliger, D. (2004) 'Ein dreistes schelmenstück, Das stromeinspeisegesetz als einfallstor für die markteinführung von erneuerbaren energien', *Solarzeitalter*, vol 1, pp12–19.

Klein, A., Held, A., Ragwitz, M., Resch, G. and Faber, T. (2006) 'Evaluation of different feed-in tariff design options', Best Practice paper for the International Feed-in Cooperation, available at www.feed-in-cooperation.org/images/files/best_practice_paper_final.pdf

Klinski, J. S. (2005) 'The Renewable Energy Sources Act (EEG) and the internal market: Compatibility of the Renewable Energy Sources Act with the current provisions of the internal electricity market and the freedom of movement of goods (expert opinion)', presented as part of the Federal Environment Ministry (BMU) project 'Legal and administrative obstacles to increasing the use of renewable energies in Germany', 24 August 2005, Berlin, available at www.wind energie.de/fileadmin/dokumente/Hintergrundpapiere/hintergrund_english/051018_Gutachten_Klinski_Binnenmarkt_en.pdf

Kofoed-Wiuff, A., Sandholt, K. and Marcus-Møller, C. (2006) *Renewable Energy Technology Deployment – RETD: Barriers, Challenges and Opportunities*, International Energy Agency (IEA), RETD, Paris

Lauber, V. and Mez, L. (2004) 'Three decades of renewable electricity policies in Germany', *Energy and Environment*, vol 15, no 4, pp599–623, available at www.ontario-sea.org/ARTs/Germany/Three%20decades%20of%20renewable%20electricity%20policy%20in%20Germany.doc

Lewis, J. and Wiser, R. (2005) 'Fostering a renewable energy technology industry: An international comparison of wind industry policy support mechanisms', LBNL-59116, November 2005, available at http://eetd.lbl.gov/ea/ems/reports/59116.pdf

L'Express (2006) 'Japan brewer pursues "Monster Cane" ethanol dream', available at www.lexpress.mu/display_article_sup.php?news_id=74417, accessed 19 October 2006

Li, Z. (2005) 'China's renewables law', *Renewable Energy World*, available at www.earthscan.co.uk/news/article/mps/UAN/432/v/3/sp/, accessed 26 January 2007

Liu, Y. (2006) 'Behind the chilly air: Impacts of China's new wind pricing regulation', Worldwatch Institute, available at www.worldwatch.org/node/3904, accessed 26 January 2007

Lovins, A. (1989) 'The negawatt revolution – solving the CO_2 problem', keynote address by Amory Lovins at the Green Energy Conference, September 14–17, 1989, Montreal, available at www.ccnr.org/amory.html, accessed 26 January 2007

Mancini, T. and Geyer, M. (2006) 'Analysis: Spain leads grid-connected solar power', available at www.euractiv.com/en/energy/analysis-spain-leads-grid-connected-solar-power/article-158769, accessed 27 January 2007

Martinot, E., Chaurey, A., Law, D., Moreira, J. R. and Wamukonya, N. (2002) 'Renewable energy markets in developing countries', *Annual Review of Energy and the Environment*, vol 27, pp309–348, available at www.martinot.info/Martinot_et_al_AR27.pdf, accessed 26 January 2007

Martinot, E., Wiser, R. and Hamrin, J. (2005) 'Renewable energy policies and markets in the United States', available at www.resource-solutions.org/lib/librarypdfs/IntPolicy-RE.policies.markets.US.pdf, accessed 20 January 2007

Methane to Markets Partnership (2006) 'About methane', available at www.methane-tomarkets.org/about/methane.htm, accessed 24 October 2006

Mez, L. (2004) 'Renewable energy policy in Germany – institutions and measures promoting a sustainable energy system', in K. Jong-dall (ed.) *Solar Cities for a Sustainable World. Proceedings of International Solar Cities Congress 2004*, Research Institute for Energy, Environment and Economy, pp528–538, available at www.eco-tax.info/downloads/EN_Mez.pdf

Ministry of Agro Industry and Fisheries (2006) 'Multi annual adaptation strategy action plan 2006–2015: Safeguarding the future through consensus', available at www.gov.mu/portal/sites/moasite/download/Multi%20Annual%20Adaption%20Strategy.pdf, accessed 26 January 2007

Mitchell, C., Bauknecht, D. and Connora P. (2003a) 'Effectiveness through risk reduction: A comparison of the renewable obligation in England and Wales and the feed-in system in Germany', *Energy Policy*, vol 34, no 3, pp297–305

Mitchell, C., Bauknecht, D. and Connora P. (2003b) 'Risk, innovation and market rules: A comparison of the renewable obligation in England and Wales and the feed-in system in Germany', WBS Centre for Management under Regulation (CMUR) Discussion Paper, available at http://users.wbs.ac.uk/group/cmur

Morris, C. (2006) *Energy Switch: Proven Solutions for a Renewable Future*, New Society Publishers, Gabriola Island, Canada

National Renewable Energy Laboratory (2004) 'Renewable energy in China: Township electrification program', available at www.nrel.gov/docs/fy04osti/35788.pdf, accessed 26 January 2007

Newsnight (2005) 'Where the car is not king', available at http://news.bbc.co.uk/1/hi/programmes/newsnight/4794361.stm, accessed 26 January 2007

Ocean Power Delivery (2006) 'The resource', available at www.oceanpd.com/Resource/default.html, accessed 20 October 2006

PassivHausUK (2007) available at www.passivhaus.org.uk/, accessed 26 January 2007

People's Daily Online (2005) 'SW province capital accelerates development of circular economy', available at http://english.people.com.cn/200501/26/eng20050126_171977.html, accessed 26 January 2007

Planet Ark (2006a) 'German firm to make biodiesel from Big Mac fat', available at www.planetark.com/dailynewsstory.cfm/newsid/39595/story.htm, accessed 22 December 2006

Planet Ark (2006b) 'North American clean technology spending hits record', available at www.planetark.com/dailynewsstory.cfm/newsid/38654/story.htm

Ragwitz, M. and Huber, C. (2004) 'Feed-in systems in Germany and Spain: A comparison', Fraunhofer Institute: Systems and Innovation Research, available at www.erneuerbare-energien.de/files/english/renewable_energy/downloads/application/pdf/langfassung_einspeisesysteme_en.pdf

Ragwitz, M., Resch, G., Faber, T. and Huber, C. (2005) 'Monitoring and evaluation of policy instruments to support renewable electricity in EU member states', Fraunhofer Institute: Systems and Innovation Research, available at www.bmu.de/files/erneuerbare_energien/downloads/application/pdf/isi_zwischenbericht.pdf

REACT (Renewable Energy Action) (2004) 'Renewable energy law, case study #7, 21 October 2004', available at www.senternovem.nl/mmfiles/Renewable%20Energy%law_tcm24-117012.pdf

Reiche, D. (ed) (2005) Handbook of Renewable Energies in European Union: Case Studies of the EU-15 States, Peter Lang, Frankfurt

REN21 (2006) 'Renewables Global Status Report – 2006 Update', available at www.ren21.net/globalstatusreport/download/RE_GSR_2006_Update.pdf

RenewableEnergyAccess.com (2006a) 'Solar cell breaks the 40% efficiency barrier', available at www.renewableenergyaccess.com/rea/news/story?id=46765, accessed 07 December 2007

RenewableEnergyAccess.com (2006b) 'Survey finds solar energy growth expected', available at www.renewableenergyaccess.com/rea/news/story?id=45350, accessed 07 July 2006

Resch, G., Faber, T., Haas, R., Ragwitz, M., Held, A. and Konstantinaviciute, I. (2006) 'OPTRES: Assessment and optimisation of renewable support schemes in the European electricity market. Potentials and cost for renewable electricity in Europe – The Green-X database on dynamic cost-resource curves', Report (D4) of the IEE project, available at www.optres.fhg.de/results/Potentials%20and%20cost%20for%20RES-E%20in%20Europe%20(OPTRES%20-%20D4).pdf

Rickerson, W. and Zytaruk, M. (2006) 'The emergence of renewable energy tariff policies in the United States', Proceedings of WINDPOWER 2006, Pittsburgh, PA, available at http://wind-works.org/FeedLaws/USA/Rickerson%20Zytaruk%20AWEA%20Windpower%202006%20presentation.pdf

Sawin, J. (2004) 'National policy instruments: Policy lessons for the advancement and diffusion of renewable energy technologies around the world', thematic background paper, Worldwatch Institute, Washington DC

Scheer, H (2005a) 'On the future of national support for renewable energy in Europe', *Photon International*, February, p80

Scheer, H. (2005b) *The Solar Economy: Renewable Energy for a Sustainable Energy Future*, Earthscan, London

Scheer, H. (2007) *Energy Autonomy: The Economic, Social and Technological Case for Renewables*, Earthscan, London

Scheuer, S. (ed) (2005) *EU Environmental Policy Handbook: A Critical Analysis of EU Environmental Legislation*, European Environmental Bureau, Brussels

Stauffer, N. (2006) 'Deep-sea oil rigs inspire MIT designs for giant wind turbines', available at http://web.mit.edu/newsoffice/2006/wind.html, accessed 2 October 2006

Stern, N. (2006) *Stern Review: Report on The Economics of Climate Change*, Cambridge University Press, Cambridge

Stryi-Hipp, G. (2006) 'Photovoltaics and solar thermal energy in Germany: Market development, applications, industry and technology', German Solar Industry Association (BSW – Bundesverband Solarwirtschaft) presentation, available at www.exportinitiative.de/media/article005926/StryiHipp_englisch_15062006.pdf

Stübner, H. (2004) 'The German Renewable Energy Act – lessons learned', InWent-Forum at the World Wind Energy Conference 2004, presentation

Sun, X. (2006) 'Wind power pricing mechanism needs overhaul', available at www.renewableenergyaccess.com/rea/news/chinawatch/story?id=46960, accessed 26 January 2007

Szarka, J. and Blühdorn, I. (2006) 'Wind power in Britain and Germany: Explaining contrasting development paths', Anglo-German Foundation, London

Teske, S. et al, Greenpeace International and European Renewable Energy Council (EREC) (2007) 'Energy [r]evolution: A sustainable world energy outlook', Greenpeace International, EREC, available at http://news.bbc.co.uk/nol/shared/bsp/hi/pdfs/25_01_07_energy_revolution_report.pdf

Thomas, J. (2006) 'The world's first 'magnetic levitation' wind turbines unveiled in China', available at www.treehugger.com/files/2006/07/china_unveils_w.php, accessed 8 October 2006

Toke, D. and Marsh, D. (2006) 'Will planners tilt towards windmills?', Report of Sustainable Technologies Programme Research Project into planning and financial issues surrounding wind power, Economic and Social Research Council, London, available at www.sustainabletechnologies.ac.uk/final%20pdf/Project%20Innovation%20Briefs/Innovation%20Brief%201.pdf

UNDESA (2005) 'Increasing global renewable energy market share: Recent trends and perspectives', prepared for Beijing International Renewable Energy Conference 2005, available at www.un.org/esa/sustdev/sdissues/energy/publications&reports/background_report_birec2005.pdf, accessed 26 January 2007

Union of Concerned Scientists (2005a) 'Renewing America's economy', available at www.ucsusa.org/clean_energy/renewable_energy_basics/renewing-americas-economy.html

Union of Concerned Scientists (2005b) 'Renewable energy tax credit saved once again, but boom-bust cycle in wind industry continues', available at www.ucsusa.org/clean_

energy/clean_energy_policies/production-tax-credit-for-renewable-energy.html, accessed 12 December 2006

Van der Linden et al (2005) 'Review of international experience with renewable energy obligation support mechanisms', LBNL-57666, available at http://eetd.lbl.gov/ea/ems/reports/57666.pdf

Veragoo, D. (2003) 'Cogeneration: The promotion of renewable energy and efficiency in Mauritius', prepared for Eastern Africa Renewable Energy and Efficiency Partnership (REEEP) Regional Consultation Meeting, available at www.reeep.org/media/downloadable_documents/Mauritius%20Country%20Paper.pdf, accessed 26 January 2007

Viertl, C. (2005) 'Political development of wind energy in Germany', presentation, Federal Ministry for Environment, Nature Conservation and Nuclear Safety, available at www.exportinitiative.de/media/article005737/BMU_Vortrag_Viertl_Frankreich_Wind_19102005_englisch.pdf

Volkery, A. and Jacob, K. (2003) 'Pioneers in environmental policy-making', Report of the Colloquium, Environmental Policy Research Centre, 18-19 October 2002, Berlin, FFU-report 05-2003, available at http://web.fu-berlin.de/ffu/download/rep_2003-05.pdf

Wiser, R. and Langniss, O, (2001) 'The Renewables Portfolio Standard in Texas: An early assessment', LBNL-49107, available at http://eetd.lbl.gov/EA/EMP/emp-pubsall.html

Wiser, R., Porter, K. and Grace, R. (2004) 'Evaluating experience with Renewables Portfolio Standards in the United States', available at http://eetd.lbl.gov/ea/ems/reports/54439.pdf, accessed 14 December 2006

World Health Organization (2005) 'Fact Sheet No 292: Indoor air pollution and health, available at www.who.int/mediacentre/factsheets/fs292/en/index.html, accessed 26 January 2007

Yang, J. (2006) 'China speeds up renewable energy development', Worldwatch Institute, available at www.frankhaugwitz.info/doks/policy/2006_10_26_China_Energy_RE_Speeds_up_RE_Development.pdf, accessed 26 January 2007

WEBSITES

Bigpicture.tv
Earthwireuk.org
Environmental-finance.com
Euractiv.com
Eurosolar.com
Martinot.info
Planetark.com
REEEP.org (Renewable Energy and Energy Efficiency Partnership)
RenewableEnergyAccess.com
UCSUSA.org (Union of Concerned Scientists)
Wind-energie.de (German Wind Energy Association)
Wind-works.org

Online Databases

DSIRE – US RE and EE database	www.dsireusa.org
REEGLE – REEEP's RE and EE database	www.reegle.info
IEA renewable energy policies and measures database	www.iea.org/textbase/pamsdb/ grindex.aspx
E-parliament – ideas and policies database	www.e-parl.net/eparliament/ ideasAction.do?action=home

INDEX

Page numbers in *italics* refer to figures, tables and boxes

in developing countries 79, 83–4
environmental issues 5
financial issues 12–13, *91*
floating turbines 115–16
net metering 15
support compared *16*
technological advances 114–16

transmission access 5–6
see also under countries named in case
 studies
Woking, Surrey 20
World Council for Renewable Energy
 (WCRE) 109